INTERNET THEORY, TECHNOLOGY AND APPLICATIONS

MAXIMIZING THE WIRELESS SPECTRUM FOR ECONOMIC GROWTH

INTERNET THEORY, TECHNOLOGY AND APPLICATIONS

Additional books in this series can be found on Nova's website under the Series tab.

Additional E-books in this series can be found on Nova's website under the E-book tab.

ECONOMIC ISSUES, PROBLEMS AND PERSPECTIVES

Additional books in this series can be found on Nova's website under the Series tab.

Additional E-books in this series can be found on Nova's website under the E-book tab.

INTERNET THEORY, TECHNOLOGY AND APPLICATIONS

MAXIMIZING THE WIRELESS SPECTRUM FOR ECONOMIC GROWTH

MATTHEW ECHOLS
AND
JANINE QUIGLEY
EDITORS

nova publishers
New York

For permission to use material from this book please contact us:
Telephone 631-231-7269; Fax 631-231-8175
Web Site: http://www.novapublishers.com

Additional color graphics may be available in the e-book version of this book.

Library of Congress Cataloging-in-Publication Data

ISBN: 978-1-62257-941-9

Published by Nova Science Publishers, Inc. † New York

CONTENTS

PREFACE

America's future competitiveness and global technology leadership depend, in part, upon the availability of additional spectrum. The world is going wireless, and we must not fall behind. The resurgence of American productivity growth that started in the 1990s largely reflects investments by American companies, the public sector, and citizens in the new communications technologies that are what we know today as the Internet. The Internet, as vital infrastructure, has become central to the daily economic life of almost every American by creating unprecedented opportunities for small businesses and individual entrepreneurs. We are now beginning the next transformation in information technology: the wireless broadband revolution. This book explores and examines maximizing the wireless spectrum for economic growth in the new era of global technology leadership.

Chapter 1 – The growth of wireless technology in the past few years has been nothing short of astonishing. The advent of smartphones, tablets, and many other devices has made mobile information access a central feature of our lives. In the coming years, access to spectrum will be an increasingly important foundation for America's economic growth and technological leadership.

In 2011, global mobile data more than doubled for the fourth year in a row. The number of devices connected to mobile networks worldwide is around five billion today, and could rise to 50 billion by 2020. By that time, wireless technologies are expected to contribute $4.5 trillion to the global economy through the expansion of existing business and the creation of new opportunities. This growth has created unprecedented demand for commercial access to wireless spectrum. At the same time, U.S. Federal spectrum needs are also rising. For example, the number of unmanned aerial systems (UAS) operated by the Department of Defense (DOD) has drastically increased from 167 to nearly 7500 from 2002 to 2010, and the systems are carrying larger payloads and collecting increased volumes of intelligence, surveillance, and reconnaissance (ISR) data. This has resulted in a dramatic increase in the number of sorties flown and domestic training requirements, all of which require spectrum.

To enhance U.S. economic competitiveness, create jobs, improve the quality of Americans' lives, and provide an environment where innovation thrives and new capabilities are secure and trustworthy, President Obama issued a Presidential Memorandum in 2010 entitled "Unleashing the Wireless Broadband Revolution" requiring that the Federal Government make available 500 MHz of Federal or non-federal spectrum for both mobile and fixed wireless broadband use by commercial users within 10 years. This President's Council of Advisors on Science and Technology (PCAST) study responds to the policy challenges and technological opportunities that have occurred in the two years since this memorandum was signed.

Chapter 2 – Presidential Memorandum: Unleashing the Wireless Broadband Revolution Memorandum for the Heads of Executive Departments and Agencies.

In: Maximizing the Wireless Spectrum ...
Editors: Matthew Echols and Janine Quigley
ISBN: 978-1-62257-941-9

Chapter 1

REALIZING THE FULL POTENTIAL OF GOVERNMENT-HELD SPECTRUM TO SPUR ECONOMIC GROWTH*

President's Council of Advisors on Science and Technology

ABOUT THE PRESIDENT'S COUNCIL OF ADVISORS ON SCIENCE AND TECHNOLOGY

The President's Council of Advisors on Science and Technology (PCAST) is an advisory group of the nation's leading scientists and engineers, appointed by the President to augment the science and technology advice available to him from inside the White House and from cabinet departments and other Federal agencies. PCAST is consulted about and often makes policy recommendations concerning the full range of issues where understandings from the domains of science, technology, and innovation bear potentially on the policy choices before the President. For more information about PCAST, see www.whitehouse.gov/ostp/pcast.

EXECUTIVE OFFICE OF THE PRESIDENT PRESIDENT'S COUNCIL OF ADVISORS ON SCIENCE AND TECHNOLOGY

President Barack Obama
The White House
Washington, DC 20502

Dear Mr. President,

We are pleased to send you this report, *Realizing the Full Potential of Government-Held Spectrum to Spur Economic Growth.* As you recognized in "Unleashing the Wireless

* This is an edited, reformatted and augmented version of President's Council of Advisors on Science and Technology Report, dated July 2012.

Broadband Revolution," your 2010 Presidential Memorandum requiring 500 MHz of spectrum to be made available for commercial use within 10 years, it is imperative that we make enough wireless spectrum available to meet the needs of rapidly expanding and innovative sectors of the economy, while also guaranteeing that the national security and public safety sectors have the spectrum they need to maintain and advance their missions.

In just two years, the astonishing growth of mobile information technology–exemplified by smartphones, tablets, and many other devices–has only made the demands on access to spectrum more urgent. This report by the President's Council of Advisors on Science and Technology (PCAST) responds to the challenges and opportunities that have arisen since your earlier Memorandum was issued. It concludes that the traditional practice of clearing government-held spectrum of Federal users and auctioning it for commercial use is not sustainable. In light of changes made possible by modern technology, we recommend that you issue a new Memorandum that states it is the policy of the U.S. government to share underutilized spectrum to the maximum extent consistent with the Federal mission, and requires the Secretary of Commerce to identify 1,000 MHz of Federal spectrum in which to implement shared-use spectrum pilot projects.

Our report, which is informed by the deliberations of PCAST members and prominent spectrum experts from the public and private sectors, identifies actions that we think the Memorandum should include so that this vision is reached, enabling multiple users to share spectrum, under a wide range of conditions, without infringing on each other's services.

To make a start on the substantial changes that PCAST proposes, the report recommends formation of an Executive Office of the President Spectrum Management Team (SMT), led by the White House Chief Technology Officer, to work with the National Telecommunications and Information Administration (NTIA) on carrying out the President's Directive. In particular, the SMT should create an accounting and incentive system to promote more effective Federal spectrum use. PCAST also recommends beginning a pilot program involving spectrum sharing, supported by early release of funds from various sources, with three key elements: immediate sharing by new low-power devices in two existing Federal spectrum bands; formation of a Spectrum Sharing Partnership Steering Committee (SSP) of industry executives (e.g. CEOs) to advise on a policy framework to maximize commercial success; and creation of an urban Test City and a Mobile Test Service that can support rapid learning in spectrum management technology and practice.

Demonstrating that the United States can move quickly to create easier access to spectrum will not only spur the domestic economy, but will help us maintain international leadership in this crucial area of modern technological innovation and commerce.

Sincerely,

John P. Holdren
Co-Chair

Eric Lander
Co-Chair

EXECUTIVE SUMMARY

The growth of wireless technology in the past few years has been nothing short of astonishing. The advent of smartphones, tablets, and many other devices has made mobile information access a central feature of our lives. In the coming years, access to spectrum will be an increasingly important foundation for America's economic growth and technological leadership.

In 2011, global mobile data more than doubled for the fourth year in a row.[1] The number of devices connected to mobile networks worldwide is around five billion today, and could rise to 50 billion by 2020.[2] By that time, wireless technologies are expected to contribute $4.5 trillion to the global economy through the expansion of existing business and the creation of new opportunities.[3] This growth has created unprecedented demand for commercial access to wireless spectrum. At the same time, U.S. Federal spectrum needs are also rising. For example, the number of unmanned aerial systems (UAS) operated by the Department of Defense (DOD) has drastically increased from 167 to nearly 7500 from 2002 to 2010, and the systems are carrying larger payloads and collecting increased volumes of intelligence, surveillance, and reconnaissance (ISR) data.[4] This has resulted in a dramatic increase in the number of sorties flown and domestic training requirements, all of which require spectrum.

To enhance U.S. economic competitiveness, create jobs, improve the quality of Americans' lives, and provide an environment where innovation thrives and new capabilities are secure and trustworthy, President Obama issued a Presidential Memorandum in 2010 entitled "Unleashing the Wireless Broadband Revolution" requiring that the Federal Government make available 500 MHz of Federal or non-federal spectrum for both mobile and fixed wireless broadband use by commercial users within 10 years.[5] This President's Council of Advisors on Science and Technology (PCAST) study responds to the policy challenges and technological opportunities that have occurred in the two years since this memorandum was signed.

PCAST finds that clearing and reallocation of Federal spectrum is not a sustainable basis for spectrum policy due to the high cost, lengthy time to implement, and disruption to the Federal mission. Further, although some have proclaimed that clearing and reallocation will result in significant net revenue to the government, we do not anticipate that will be the case for Federal spectrum. In March of 2012, the National Telecommunications and Information Administration (NTIA) concluded that clearing just one 95 MHz band by relocating existing Federal users to other parts of the spectrum would take 10 years, cost some $18 billion, and cause significant disruption to incumbent users.[6] The last successful auction that involved cleared Federal spectrum, in 2006, yielded a total of $13.7 billion for 90 MHz, but only half of the auctioned spectrum was Federal (the other half was already commercial), and the Federal agencies then required $1.5 billion over the next 6 years to relocate services out of the cleared bands. In the end, therefore, the Federal contribution of 45 MHz realized a net of just $5.35 billion.[7] When this net revenue is annualized over 10 years or more, the typical duration of a license, the amount of revenue the Federal Government will receive is small. These modest sums should not be driving the direction of spectrum policy.

Historically, spectrum was managed by assigning exclusive rights to use a specific frequency in a specific location. Initially, these authorizations were granted to governmental and commercial users at no cost. Since the mid-1990s, long term commercial licenses have generally been assigned through competitive auctions. Winning bidders typically receive spectrum access in the form of exclusive assignments of frequencies to chosen services (i.e., licenses), ensuring that no other services infringe on that assignment (i.e., no interference). This study finds that today's apparent shortage of spectrum is in fact an illusion brought about because of the way spectrum is managed. If the Nation instead expands its options for managing Federal spectrum, we can transform the availability of a precious national resource —spectrum—from scarcity to abundance. This expansion can be done in such a way that it will not result in a loss of revenue to the Federal Government and may result in new revenue either from enhanced economic growth and innovation or from modest leasing fees. But in either case, the value to the Federal Government will be greater if the spectrum is available for reuse or relicensing more often than it is today. The new system for Federal spectrum management that this report calls for—a new spectrum architecture and a corresponding shift in the architecture of future radio systems that use it—can multiply the effective capacity of spectrum by a factor of 1,000.[8]

The essential element of this new Federal spectrum architecture is that the norm for spectrum use should be sharing, not exclusivity. Technology innovations of recent years make this transformation eminently achievable. Two trends are especially important. First, instead of just the tall cell towers that provide coverage for very large geographic areas, many wireless services are already moving to "small cell" operations that provide services for very small geographic areas, reducing the potential for interference so that other services may operate much closer to them. The huge explosion of Wi-Fi services is one example of this evolution. Second, improvements in performance make it possible for devices to deliver services seamlessly even in the presence of signals from other systems, so that they do not need exclusive frequency assignments, only an assurance that potentially interfering signals will not rise above a certain level.

Taking these and other developments into account, this report argues that spectrum should be managed not by fragmenting it into ever more finely divided exclusive frequency assignments, but by specifying large frequency bands that can accommodate a wide variety of compatible uses and new technologies that are more efficient with larger blocks of spectrum.

The recommendations in this report are based on starting with low-risk existing technologies, early versions of which are already being deployed today. Enacting these recommendations will create market opportunity for newer technologies, enabling them to mature faster, accelerating the growth of spectrum sharing capacity, and leading to the development of an ongoing innovation cycle. However, the policies proposed are consistent with the later deployment of these non-commercial technologies, only when they are validated for their operational use in Federal spectrum.

To make an analogy, today's spectrum use resembles road transportation at the beginning of the automotive revolution when we created our highways and interconnection and commerce flowed. The mid-1980s innovation of "unlicensed" spectrum use, which makes spectrum available at no cost to any user willing to abide by technical conditions of use, has been essential to the rise of Wi-Fi and represents a wireless analogy to the early shared roadways. The rest of the spectrum system, however, still looks like a series of narrow roads. What PCAST proposes is creating the spectrum equivalent of wide multi-lane superhighways, where the lanes are continuously shared by many cars, trucks, and other vehicles. Spectrum superhighways would be large stretches of spectrum that can be shared by many different types of wireless services, just as vehicles share a superhighway by moving from one lane to another. In contrast to the way we have allocated spectrum, the road system has always let Federal and commercial vehicles share the same highways, with the proviso that government use was allowed to preempt commercial users' rights for reasons of public safety, emergency medical rescue, or national security. There is no reason that the same principles cannot apply to spectrum management. Users of spectrum can make use of the wireless equivalents of signals, sensors, and stop lights to avoid "collisions" with other users. Just as we created the initial transcontinental superhighways in the 20-30 years that followed the 1939 FDR-commissioned blueprint "Toll Roads and Free Roads,"[9] we have the chance to create spectrum superhighways today.

As a result, the most urgent recommendation in this report is that the President issue a new memorandum that states it is the policy of the U.S. government to share underutilized Federal spectrum to the maximum extent possible that is consistent with the Federal mission, and requires the Secretary of Commerce to immediately identify 1,000 MHz of Federal spectrum in which to implement the new architecture and thereby create the first shared-use spectrum superhighways. Taking this step represents a continuation and an expansion of the President's Directive of 2010.

As part of the process to reach this 1,000 MHz goal,[10] PCAST recommends that the Federal Government, using industry partners, establish a new Federal Spectrum Access System (SAS) that will serve as an information and control clearinghouse for band-by-band spectrum registrations and conditions of use and allow non-Federal users to access underutilized spectrum in Federal bands. The SAS will put into practice the fundamental principle that underutilized spectrum capacity should be used or shared to the greatest possible extent. Another recommended change is that Federal spectrum, instead of being divided into small, dedicated frequency blocks as it is at present, should be divided into substantial frequency blocks spanning several hundred megahertz. Establishing these wide bands will make it easier for spectrum sharing to be the norm, a transformation in which all Federal agencies would be required to cooperate. Making spectrum access available to a wide range of services and applications will also require provision of a framework that establishes minimum technical standards for the coexistence of transmitters and receivers, in contrast to the present system that focuses on transmitters. Finally, simple measures that assess individual spectrum uses solely by their need for megahertz must be replaced by more sophisticated metrics that reveal how effectively a stretch of spectrum can accommodate a variety of complementary services within a given area.

We recognize that the new spectrum architecture proposed in this report represents a major evolution of existing spectrum management practices. Implementing it will not be easy and may take a long time. But just as the transcontinental highway system began with one road, we must act immediately to act on the initial 1,000 MHz. Before they will embrace the new system, incumbent Federal spectrum users will need to have confidence that sharing of the spectrum they have been allocated will not cause harmful interference to the technologies that they operate, and commercial operators with new technologies will need to be made sure of the reliability of the spectrum access needed for their business models. So, to get started, we are proposing three key elements of a significant pilot program that includes immediate actions toward implementing our recommendations:

1. The immediate sharing of new low-power civil devices in two existing Federal bands, of over 100 MHz combined.

2. The creation of a group of industry executives (e.g. CEOs), selected by the President and called the Spectrum Sharing Partnership Steering Committee (SSP), to recommend a policy framework, centered on a public private partnership for sharing Federally-held spectrum, and implementation milestones that lay the groundwork for the first spectrum superhighways. We expect the SSP to make its recommendations over a one-year timeframe as opposed to being a longterm ongoing effort As necessary, they may wish to call upon the NTIA Commerce Spectrum Management Advisory Committee (CSMAC) and the Federal Communications Commission (FCC) Technology Advisory Council (TAC) for technical advice. The products might be short memorandum or reports focused on specific topics the SSP believes are important to address for the effort to be successful.
3. The creation of an urban Test City in a major U.S. city along with a Mobile Test Service that can relocate to urban, rural, and Federal facilities as needed to support rapid experimentation in spectrum management technology and practice.

We estimate that the overall costs of implementing this program, over the next 3 years, will be in the range of about $80 million. We view the Federal Government as the initial funding source to cover costs, along with a public private partnership that will have the aim of transferring most costs to the private sector over the course of time (see Table ES.2).

We believe this shift in direction will also require increased White House involvement. Specifically, we recommend that that the White House Chief Technology Officer (CTO), with equivalent level representatives from the National Security Staff (NSS), the Office of Management and Budget (OMB), and National Economic Council (NEC) formalize a Spectrum Management Team (SMT) to work with the NTIA to carry out the President's directive.

Federal users currently have no incentives to improve the efficiency with which they use their own spectrum allocation, nor does the Federal system as a whole have incentives to improve its overall efficiency.

This report therefore proposes an accounting, allocation, and incentive system (nominally called "Spectrum Currency") that would work in conjunction with a Spectrum Efficiency Fund (our recommended evolution of the current Spectrum Relocation Fund), administered by the OMB, to reward agencies that move quickly to promote more effective spectrum use by making some of their spectrum available for sharing with other Federal and non-Federal users (see Table ES.2).

One of the other important directions that spectrum policy must take is to create a marketplace that can accommodate the widest range of commercial users, from initial venture-funded startups to established service providers. Today's spectrum ecosystem offers only the choice between unlicensed and long term, renewable licensed spectrum. The number of business entities that can participate in auctions for nationwide, long term spectrum licenses, is limited. Experimenting with shorter-term, lower cost, spectrum license options for commercial users sharing Federal spectrum, will foster new innovative ideas, increase the number of participants in this market, contribute to economic growth, and also provide a way to collect an ongoing stream of revenue, if that is desired.

Table ES.1. PCAST Spectrum Report Findings and Recommendations

Topic	Findings	Recommendations
From Scarcity to Abundance	Finding 1.1: Spectrum provides a great economic opportunity for the Nation. The economy created by making spectrum abundant has the opportunity to provide social benefits of over $1 trillion and millions of jobs for Americans over many years. Most importantly, it will provide a foundation for American economic and technological leadership. Finding 1.2:Clearing and reallocation of Federal spectrum for exclusive use is not a sustainable basis for spectrum policy due to the high cost, lengthy time to implement, and disruption to the Federal mission. Sharing of Federal spectrum, however, would provide the basis for economic and social benefits for the Nation. Finding 1.3:The fragmented partitioning of Federal spectrum leads to inefficiency, artificial scarcity, and constraints on current and future Federal and non-Federal uses.	Recommendation 1.1: PCAST recommends that the President issue a new memorandum that states it is the policy of the U.S. government to share underutilized Federal spectrum to the maximum extent possible that is consistent with the Federal mission, and requires the Secretary of Commerce to immediately identify 1,000 MHz of Federal spectrum in which to implement the new architecture and thereby createthe first shared-use spectrum superhighways.
Policy Hurdles to Clear	Finding 2.1:Sharing of Federal spectrum provides an opportunity to deploy a wholly new approach to Federal spectrum architecture and policy by establishing large shared spectrum blocks, new effectiveness metrics, and coordinated and prioritized Federal and commercial use.	Recommendation 2.1:The Secretary of Commerce, in collaboration with the Federal Communications Commission (FCC), should establish a mechanism to provide the Federal Government with the ability to manage the sharing of Federal spectrum. Federal spectrum should be divided into substantial frequency blocks with common characteristics, rather than the current narrow band servicespecific static allocation scheme. In addition, rather than the current preallocation and assignment of spectrum, there should be a new"dynamic sharing" model that makes spectrum sharing by Federal users the norm, and also allows sharing with commercial users.
	Finding 2.2: Wireless architectures have evolved from a single model of high-power, high altitude base stations to a mix of capabilities, ranging all of the way from base stations to offload onto commercial Wi-Fi. This provides an opportunity to locally exploit Federal spectrum sharing opportunities that would not be otherwise compatible with high power operations (such as LTE).	Shared access to Federal spectrum should be governed according to a threetier hierarchy: Federal primary systems would receive the highest priority and protection from harmful interference; secondary licensees must register deployments and use in a database and may receive some quality of service protections, possibly in exchange for fees; and General Authorized Access users would be allowed opportunistic access to unoccupied spectrum to the extent that no Federal Primary or Secondary Access users are actually using a given frequency band in a specific geographical area or time period. All Federal agencies should be required to cooperate in the implementation of these changes.
		Recommendation 2.2:The Secretary of Commerce, working through the National Telecommunications and Information Administration (NTIA) and the FCC, should authorize and implement, directly or through commercial providers, a Federal Spectrum Access System (SAS) to serve as an information and control clearinghouse for the band-by-band registrations and conditions of use that will apply to all users with access to each shared Federal band under its jurisdiction.

Table ES.1. (Continued)

Topic	Findings	Recommendations
		The SAS will protect Federal operations from interference while allowing nonFederal users to access underutilized spectrum in Federal bands. Underutilized spectrum capacity in Federal bands should be made available to the greatest possible extent for noninterfering shared use, based on the principle that exclusive assignments should not be taken as a justification for letting unused or underutilized spectrum lie fallow.
Technology Advancements and Challenges to Solve	Finding 3.1:Spectrum management and regulation is focused on the characteristics of transmission, whereas receiver characteristics increasingly constrain effective and flexible spectrum usage	Recommendation 3.1:The Secretary of Commerce working through the National Telecommunications and Information Administration (NTIA), in cooperation with the Federal Communications Commission (FCC), should establish method-ologies for spectrum management that consider both transmitter and receiver characteristics to enable flexible sharing of spectrum. To safeguard primary Federal users, FCC should require that future nonFederal devices will be permitted to share government spectrum as Secondary Access users only if they are certified to operate within the stated interference limits for the band of interest. Initial specification of protection should be reviewed such that they safeguard new FCC assignments against harmful interference while grandfathering in existing devices andoperations.
New Application Economy	Finding 4.1:Moving to a dynamic sharing model for Federal spectrum would unlock economic benefits by allowing the private sector to make intensive use of currently underutilized parts of the radio spectrum. A welldesigned Federal spectrum policy opens up opportunities for innovation and growth in sectors that are barely imagined, much less welldefined, when the policy choice is made. Finding 4.2: Sharing of Federal spectrum provides an opportunity to deploy new spectrum management principlessuch as shorter term licenses that would be appropriate to new and innovative spectrumbased services and products. This provides an opportunity to collect revenue to the Treasury from the private sector for assured use of spectrum.	Recommendation 4.1: PCAST recommends that policies enabling commercial access to Federal spectrum be based primarily on their effects on innovation and growth in wireless devices, services, and associated markets; direct revenue considerations should be treated as secondary. The Office of Management and Budget (OMB) should develop a model to assess future economic growth effects of wireless allocations as well as revenue from increased economic activity. The National Telecommunications and Information Administration (NTIA) and the Federal Communications Commission (FCC) should continue to embrace the current proven unlicensed model and, for licensed spectrum, explore adding new short and mediumterm spectrum license models that could both foster growth for these new applications and collect revenue.
Starting with Federal Spectrum	Finding 5.1: There is no incentive system today for Federal G overnment agencies to be efficient in their use of spectrum or to share spectrum allocated to them with the non-Federal sector. Finding 5.2: A public private partnership (PPP) is the best mechanism to ensure that optimal use is made of the Federally-held spectrum and of related investments in spectrum research and testing.	Recommendation 5.1:PCAST recommends that the White House Chief Technology Officer (CTO) with senior officials at an equivalent level from the National Security Staff (NSS), the Office of Management and Budget (OMB), and the National Economic Council (NEC) formalize a Spectrum Management Team (SMT) to work with the National Telecommunications and Information Administration (NTIA), the Federal Communications Commission (FCC), and the major Federal agencies that use spectrum to carry out the President's directive.

Topic	Findings	Recommendations
	Finding 5.3:International harmonization of spectrum policies is essential to product innovation, interoperability and roaming, spectrum efficiency, and cross-border frequency coordination.	Recommendation 5.2:PCAST recommends that the NTIA, working with the SMT and Federal agencies, reexamine the partitioning of Federal spectrum usage in light of current and emerging technology. One objective of this reexamination is to aggregate current spectrum partitions to create substantial frequency blocks in order to facilitate sharing through common technical use rules. Recommendation 5.3:PCAST recommends that the President indicate that all Federal agencies should cooperate with the SMT and NTIA to establish and implement a government-wide process and mechanism to share Federally-held spectrum. Within one year, the SMT working with the NTIA should formulate concrete 5year and 10year goals for Federal spectrum sharing opportunities in order to recommend to the President how to appropriately update his 2010 goal of making 500 MHz of Federal and nonFederal spectrum available over the next 10 years. Recommendation 5.4: PCAST recommends that OMB, working with the SMT and NTIA, take steps to implement a mechanism that will give Federal agencies incentives to share spectrum.Such a mechanism would accurately internalize the opportunity cost of Federal spectrum resources and manage them over long time horizons using a "currencylike" accounting, allocation, and incentive system ("Spectrum Currency"). Recommendation 5.5:PCAST recommends that OMB should implement a sustainable funding mechanism to foster a Federal spectrum sharing system.The existing Spectrum Relocation Fund should be redefined as a revolving "Spectrum Efficiency Fund" that recycles private sector payments for use of Federal spectrum into reimbursements to Federal agencies for investments that facilitate spectrum sharing and enhance spectrum efficiency. Congress should allow the Fund to reimburse qualifying costs by any Federal service, not just those in revenuegenerating bands.
		Recommendation 5.6: PCAST recommends that the President appoint an advisory committee of industry executives (e.g. CEOs), to be known as the Spectrum Sharing Partnership Steering Committee (SSP), to advise the SMT on a policy framework to maximize commercial success, centered on a public private partnership for sharing Federally-held spectrum, and implementation milestones that lay thegroundwork for the first spectrum superhighways. Recommendation 5.7: The United States, represented by the Department of State with advice from NTIA and the FCC, should make international harmonization of spectrum allocations to wireless broadband, particularly in bands used or planned to be used for mobile broadband applications in the United States, a key element of the U.S. position at the 2015 World Radiocommunication Conference (WRC-15) and in bilateral and regional discussions with its own neighbors, Mexico and Canada.

Table ES.1. (Continued)

Topic	Findings	Recommendations
Implementation	Finding 6.1:Insufficient opportunities are available to test new architectures, policies, and the new systems proposed in this reportfor the large scale dynamic sharing of innovative commercial products in the presence of existing real world public safety and Federal incumbent applications.	Recommendation 6.1:PCAST recommends that the Secretary of Commerce, working through the NationalTelecommunications and Information Administration (NTIA) and the National Institute of Standards and Technology (NIST), provide test services (a Test City and a related Mobile Test Service) to support the development of the policies, underlying technologies, and system capabilities required to support dynamic spectrum sharing. Services would include largescale sustainable facilities for systemslevel testing across multiple frequency bands, including public safety and selected Federal bands. The Secretary should support these services by establishing a Public Private Partnership (PPP) that would pool the resources of Federal, state, and local governments with industry and academia. The Federal contribution to the partnership could be funded, depending on timing and other factors, by NIST's Wireless Innovation Fund, by the Public Safety Trust Fund, and potentially by the Office of the Secretary of Defense and the National Science Foundation.

Table ES.2. Proposed PCAST Spectrum Federal Funding Mechanisms

Ongoing agency funding mechanisms for agency equipment monitored by OMB		
Funding Mechanism	Purpose	Why is it needed?
Spectrum Currency:Each agency would be given an allocation of a synthetic currency that they could use to "buy" their spectrum usage rights. OMB would administer this system with the initial valuations by using comparable private sector uses for which the market has already set a price.	Provide an initial economic score and then incentives to Federal agencies to be efficient in their spectrum allocation use including reducing their own need for spectrum, by sharing spectrum with other agencies and nongovernment users.	For most procurement activities, agencies are limited based on market mechanisms such as budget. This is not true for spectrum, so there are no incentives to be efficient. Reducing their use of synthetic Spectrum Currency wouldreward early adopters of improved spectrum effectiveness with a trade for real dollars from the Spectrum Efficiency Fund.
Spectrum Efficiency Fund:Each agency would receive financial resources from the proposed Spectrum Efficiency Fund (the broadened and repurposed Spectrum Relocation Fund (SRF)) to reimburse agencies for the costs of research, planning, and testing to prepare for sharing. The SRF was established by Congress in 2004 with the explicit and limited purpose of reimbursing agencies for the actual costs incurred in relocating Federal systems from auctioned bands.	Provide incentives to Federal agencies to take on the cost and risk of updating their system technology to accommodate and facilitate spectrum band sharing and to encourage them to make efficient use of spectrum.	Federal agencies may have no incentive or authority to enhance their use of spectrum if the cost depletes the budget available for their core mission. As a result, they may decide not to take on the substantial costs of relocating agency systems and operations, expanding shared access to Federal bands, designing or procuring new and upgraded Federal systems, or moving to far more spectrumefficient and/or interferencetolerant technologies.

Ongoing agency funding mechanisms for agency equipment monitored by OMB		
Funding Mechanism	Purpose	Why is it needed?
The 2012 Payroll tax agreement broadened the purposes of the SRF to also include improving Federal systems left in bands that have been auctioned for commercial use.		
Mechanisms to pay for implementation costs of the PCAST study recommendations		
Wireless Innovation Fund (WIN): This is initially a $100 million fund at the National Institutes of Standards and Technology (NIST), and is part of the 2012 Payroll tax agreement for Spectrum R&D. (It also adds an additional $200 million after $20.4 billion of auction income has been secured).	Provide funding, depending on timing and other factors, for NIST to determine if dynamic sharing can be used for public safety by funding the public safety test beds that are an integral part of the PCAST proposed Test city.	Funds are needed for PCAST's proposed Test City and Mobile Test Service with estimated costs for construction and operation for the first three years of approximately $60million.
Public Safety Trust Fund (PSTF): As required by the 2012 Payroll tax agreement, funds from the incentive auctions carried out by the FCC can be used to repay amounts borrowed by NTIA (see above). Only $7 billion is allocated in Federal funding in the PSTF for Public Switched Broadband Network (PSBN) development and operation. Ongoing PSBN operation requires further private sector funding sources	If PCAST's recommended system is implemented, after being tested in the Test City, as funded by the WIN fund, usage fees obtained from Secondary Access users of Public Safety spectrum could be collected as revenue for the PSTF.	PSTF requires future funding after the initial $7 billion to expand and operate PSBN. Proving the viability ofthe PCAST recommended system of primary and Secondary Access users sharing Public Safety spectrum, would give Public Safety Federal Primary Access users the comfort to go forward to allow sharing of their spectrum. The implemented primarysecondary system could then collect usage fees from secondary private sector Public Safety spectrumusers to replenish the PSTF.
Public Private Partnership (PPP):This is a new entity that PCAST recommends that the Secretary of Commerce form.	See above. PCAST recommends that, over time, most of the costs for running the Test City should be transferred to a PPP.	In order for the United States to be competitive in its use of spectrum, government, industry, and academia need to work together to ensure industry needs are met.

Although complete accomplishment of this transformation, in all Federal spectrum, will take time— perhaps two to three decades—we stress that implementing our recommendations will lead to rapid results. The long term direction outlined in this report can start to be operational in 1-3 years. Unless we make a determined and significant move in this direction, the United States risks falling behind other countries that are equally aware of looming problems in spectrum management and the huge advantage to be gained by solving them.

PCAST Spectrum Policy Invited Experts

PCAST Members

Mark Gorenberg (chair)
Managing Director
Hummer Winblad Venture Partners

S. James Gates, Jr.
John S. Toll Professor of Physics
Director, Center for String and Particle Theory
University of Maryland, College Park

Craig Mundie
Chief Research and Strategy Officer
Microsoft Corporation

William Press
Raymer Professor in Computer Science
and Integrative Biology
University of Texas at Austin

Maxine Savitz
Vice President
National Academy of Engineering

Eric Schmidt
Executive Chairman
Google, Inc.

Invited Experts

Yochai Benkler
Berkman Professor of Entrepreneurial Legal Studies and codirector,
Berkman Center for Internet and Society Harvard University

Jennifer Bernhard
Professor of Electrical and Computer Engineering
University of Illinois at Urbana-Champaign

Vanu Bose
President and CEO
Vanu Inc.

Michael Calabrese
Director, Wireless Future Project,
Open Technology Institute
New America Foundation

Dale Hatfield
Senior Fellow, Silicon Flatirons Center for Law,
Technology, and Entrepreneurship
University of Colorado, Boulder

Michael Katz
Sarin Chair of Strategy and Leadership
University of California, Berkeley

Paul Kolodzy
Kolodzy Consulting

William Lehr
Research Associate
Computer Science and Artificial
Intelligence Laboratory
Massachusetts Institute of Technology

Jon Levin
Holbrook Working Professor and Chair
Department of Economics
Stanford University

David Liddle
Venture Partner
U.S. Venture Partners

Preston Marshall
Deputy Director, Information Sciences Institute,
Research Professor
University of Southern California

J.D. McCreary
Chief, Strategic Technology Program Office
Sensors and Electromagnetic Applications Lab
Georgia Tech Research Institute

Mark McHenry
President, CTO, and Founder
Shared Spectrum

Milo Medin
Vice President, Access Services
Google, Inc.

Teresa Meng
Reid Weaver Dennis Professor of Electrical
Engineering
Stanford University

Jeff Reed
Willis G. Worcester Professor of Electrical and
Computer Engineering
Virginia Tech

Dennis Roberson
Vice Provost and Professor of Computer Science
Illinois Institute of Technology

Gregory Rosston
Deputy Director, Stanford Institute for
Economic Policy Research, and
Deputy Director, Public Policy Program
Stanford University

Pierre de Vries
Senior Adjunct Fellow, Silicon Flatirons Center for Law,
Technology, and Entrepreneurship
University of Colorado, Boulder

Kathleen Wallman
President
Wallman Consulting, LLC.

Tom Wheeler
Managing Director
Core Capital

Staff

Deborah Stine
Executive Director, PCAST

Danielle Evers
AAAS Science and Technology Policy Fellow

Ben Buchanan
Student Volunteer

Thomas Power
Deputy Chief Technology Officer,
Telecommunications Office of Science
and Technology Policy

David Lindley
Writer

Federal Agency Liaisons

John Leibovitz
Deputy Chief, Wireless Telecommunications Bureau
Federal Communications Commission

Douglas Sicker
Chief Technology Officer and Senior Advisor for Spectrum
National Telecommunications and
Information Administration

I. INTRODUCTION

Finding 1.1: Spectrum provides a great economic opportunity for the Nation. The economy created by making spectrum abundant has the opportunity to provide social benefits of over $1 trillion and millions of jobs for Americans over many years. Most importantly, it will provide a foundation for American economic and technological leadership.

Finding 1.2: Clearing and reallocation of Federal spectrum for exclusive use is not a sustainable basis for spectrum policy due to the high cost, lengthy time to implement, and disruption to the Federal mission. Sharing of Federal spectrum, however, would provide the basis for economic and social benefits for the Nation.

Finding 1.3: The fragmented partitioning of Federal spectrum leads to inefficiency, artificial scarcity, and constraints on current and future Federal and non-Federal uses.

> **Recommendation 1.1:** PCAST recommends that the President issue a new memorandum that states it is the policy of the U.S. government to share underutilized Federal spectrum to the maximum extent possible that is consistent with the Federal mission, and requires the Secretary of Commerce to immediately identify 1,000 MHz of Federal spectrum in which to implement the new architecture and thereby create the first shared-use spectrum superhighways.

1.1. The Need and Opportunity to Create the Spectrum Super Highway

Few developments hold as much potential as wireless technologies to enhance America's economic growth and improve our communications, business productivity, and core activities like public safety, healthcare, education, and electric utilities. In short, wireless technologies have become an indispensable element of our overall quality of life.

In 2011, global mobile data more than doubled for the fourth year in a row.[11] According to the Organisation for Economic Cooperation and Development (OECD), the number of devices connected to mobile networks worldwide is around five billion today and could rise to 50 billion by 2020.[12] The implementation of wireless technologies will generate new revenues and business models as well as enhance the efficiency and effectiveness of current services, leading to an estimated global business impact of up to $4.5 trillion by 2020.[13] And studies of earlier generations of communications technology suggest that increased penetration is strongly associated with growth. A 2009 World Bank study found that a 10% increase in broadband capacity was associated with a 1.3% increase in economic growth.[14]

Wireless networks are especially useful, compared to fixed broadband, in contributing to socio-economic value because they enable mobile access, supply connectivity at lower costs in low density areas, and eliminate the need to wire or rewire when network configurations change. So even when wireless connectivity is not the only solution, it is often the preferred solution. Most of the socio-economic value of broadband now comes from wireless media.[15] The United States has seen vigorous economic growth from the use of spectrum licensed for exclusive use, notably in nationwide cellular and wireless broadband services. A recent study of data from 2008 estimated total cellular revenue in the United States of $141 billion and consumer surplus (value to consumers above what they pay) of $212 billion.[16] Remarkable economic growth has also come from access to unlicensed regions of spectrum, which are open to opportunistic access by all users, subject to certain general restrictions and with no guarantees as to the quality of service available (see Box 1.1). Wi-Fi and Bluetooth are two well-known and hugely successful technologies that have evolved in unlicensed spectrum. One recent report estimated the value produced by unlicensed spectrum at $50 billion a year;[17] another estimated the value of Wi-Fi alone at $52-99 billion a year.[18] Table 1.1 illustrates the enormous range of applications using both licensed and unlicensed access to spectrum.

This creation of new industries and associated economic activity took place largely in the United States because it was the first country to provide the opportunity to develop and deploy the technology. This provided U.S. firms with a head start in areas such as Internet applications, equipment, and services, and unlicensed communications. But continued leadership in these new applications, especially those involving mobile data, requires the

United States to redouble its efforts to be innovative because a shortage of available spectrum will put U.S. industry at a worldwide competitive disadvantage. The Nation has been at this crossroads before in cellular communications. The United States created the mobile wireless industry, starting with the introduction of the first car radio by Motorola in 1930. Then came an early application in public safety with police radios, followed by "walkie-talkie" communications in World War II, thirteen major communications units on board each Apollo manned space mission in the late 1960s that allowed words and important telemetry data to be communicated back to Earth, and the first handheld mobile phone in 1973. But starting with the launch of the first GSM[19] system in Finland in 1991, the center of gravity for cellular-based mobile equipment research, technology and manufacturing migrated off shore. While, in the past few years, the United States once again houses leaders like Apple, Google, and Cisco in mobile phone operating systems, application store providers, and Wi-Fi routers, none of the major base station equipment vendors supplying the cellular carriers are headquartered in North America.[20,21]

Table 1.1. Variety of Spectrum Applications

Cellular telephone systems	Wireless broadband	Civil and military radar
WiFi Devices—Home and business networks: Hotspots	Tanklevel meters	Industrial automation controls
Community, urban & rural broadband	Traffic light controls	RFID systems
Bluetooth headsets & keyboards	Crane controls	Retail antitheft systems
Automobile keyless entry	Lighting controls & dimmers	Security alarm systems
Inhome video distribution	Wireless door bells	Wireless speakers
Remote control toys	Cordless phones	Satellite RadiotoFM radio
Toy walkietalkie	Garage door opener controls	Convergence w licensed devices
Utility meter readers & smart grid energy control	Sensors for automatic doors	Meat thermometers
Medical camera pills	Inventory control	Diaper wetness sensor
Medical panic alerts	Pool cover controllers	And the list goes on....

Source: Presentation by Julius Knapp, FCC, IDGA 2012 Conference.

Box 1.1. Unlicensed Spectrum Explained

Unlicensed spectrum refers to radio frequency bands in which technical rules are specified for both the hardware and deployment of radio systems that are open for shared use by an unlimited number of compliant users. Although the Federal Communications Commission (FCC) permits unlicensed use at extremely low power in most frequency bands, the FCC has adopted rules allowing operation of higher power unlicensed devices in certain bands—such as the so-called Wi-Fi band at 2.4 GHz—where there is significantly more unlicensed usage. Unlicensed devices authorized under Part 15 of the FCC's rules include cordless telephones, garage door openers, baby monitors and microwave ovens, as well as broadband networks and a rapidly evolving collection of new technologies. The term "unlicensed" is something of a misnomer since use is, in fact, regulated to ensure that unlicensed devices do not cause interference to operations with a higher priority.

Any person or entity may use unlicensed spectrum for either private or public purposes so long as the user's equipment is certified by the FCC and operated in conformity with Part 15 of the Commission's rules.

Unlike most licensed spectrum users, unlicensed spectrum users enjoy no regulatory protection against interference from other licensed or unlicensed users in the band. Although FCC device certification rules and standardized protocols (such as the Wi-Fi Alliance's 802.11 family of protocols) help to mitigate interference, users must accept any interference caused by all compliant devices in the band.

Source: Adapted from www.ntia.doc.gov/files/ntia/meetings/unlicensed spectrumsub committee report_01102011.pdf.

The competition for international leadership in the mobile space is ongoing and tenacious. President Barack Obama, in his Presidential Memorandum "Unleashing the Wireless Broadband Revolution" of June 28, 2010, stated that "America's global technology leadership, and the worldwide wireless revolution, depends on one of our greatest natural resources—wireless spectrum."[22] The economy created by making spectrum abundant has the opportunity to provide societal value of over $1 trillion and millions of jobs for Americans in the coming decade.[23,24]

But in its 2010 analysis that led to the National Broadband Plan, the Federal Communications Commission (FCC) expected the United States to see a spectrum shortage as early as 2014.[25] Insufficient spectrum could limit coverage, service quality, and data connection speeds.[26] Spectrum is also essential to the Federal Government which uses spectrum for emergency communications, national security, law enforcement, aviation, maritime, space communications, and other Federal needs (see Figure 1.1).

In response to these needs, that same Presidential Directive of 2010 provided specific direction to the FCC and National Telecommunications and Information Administration (NTIA) to obtain access to an additional 500 MHz for commercial use within 10 years.

Since the Presidential Directive, however, and even with the current FCC plans to repurpose significant commercial spectrum, the attempt to clear and reallocate spectrum, particularly Federal spectrum, to meet the 500 MHz target has been progressing slowly. A recent NTIA report on the potential for reallocation of the 1755-1850 MHz band, a high priority candidate in the search for 500 MHz, also demonstrates that full relocation of government users (see Figure 1.2) may have significant operational impact and will take up to 10 years and cost some $18 billion.[27]

These relocation costs may even exceed the likely revenue raised by auctioning the band to commercial users. Second, when the 3550–3650 MHz band was being considered last year for auction, the need to preserve essential Federal services meant that spectrum from 3550 MHz to 3650 MHz could only be offered with very large exclusion zones extending inland nearly 200 miles from both the east and west coasts of the United States and including a majority of the U.S. population.[28]

And, as the need for spectrum to support the increasingly mobile economy has grown, significant tension between different users—cellular, Federal, public safety, and unlicensed—has emerged.

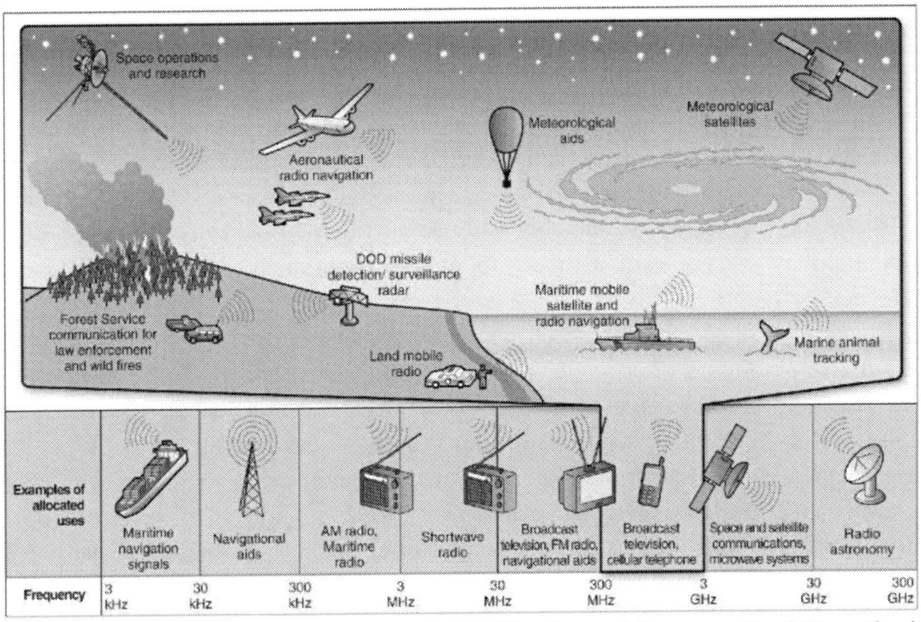

Source: GAO. (2011). Spectrum Management: NTIA Planning and Processes Need Strengthening to Promote the Efficient Use of Spectrum by Federal Agencies. www.gao.gov/new.items/d11352.pdf.

Note: GAO analysis of NTIA, federal agencies, and industry information.

Figure 1.1. Examples of Allocated Spectrum Uses, and Federal Spectrum Use in the High-Value Range.

While the United States is challenged to move ahead with its broadband plans, China, Germany, Hong Kong, and South Korea have developed broadband plans to upgrade their wireless and wire-line broadband platforms at a fast rate. And on March 7, 2012, Ed Richards, Chief Executive of the U.K. Office of Communications (Ofcom), the British equivalent of the FCC, gave a bold speech in Brussels challenging Europe to move in a new innovative direction that embraces the technologies and policies needed for worldwide leadership.[29]

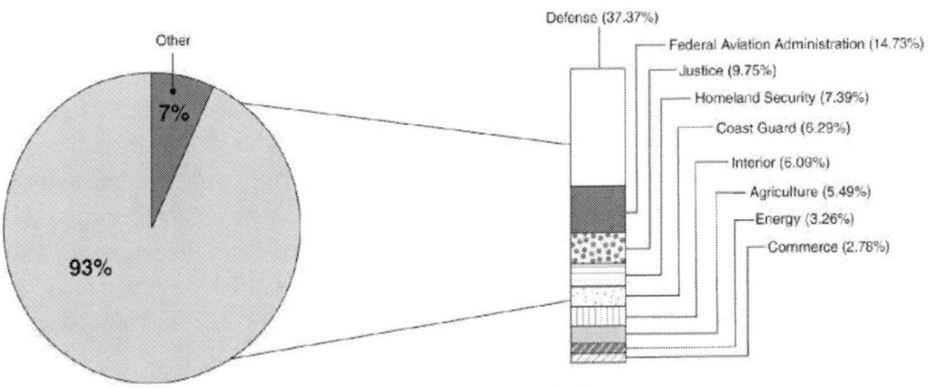

Source: GAO. (2011). Spectrum Management: NTIA Planning and Processes Need Strengthening to Promote the Efficient Use of Spectrum by Federal Agencies.www.gao.gov/new.items/d11352.pdf.

Note: GAO analysis of NTIA GMF spectrum assignment data, September 7, 2010.

Figure 1.2. Federal Agencies with the Most Spectrum Assignments.

It is a fundamental fact of physics that new spectrum cannot be manufactured. This poses particular difficulty in the regions of the spectrum that are the most in demand for current communication system architectures. However, this seemingly intractable problem presents us with an enormous opportunity. The basic model of spectrum allocation and management has not changed in 100 years. Fortunately, digital technologies have evolved so dramatically during the last 50 years, and many new spectrum technologies have become feasible in the past 10 years, that we now have the possibility of moving to a radically different model.[30] A significant policy change can be made to move spectrum availability from scarcity to abundance.

To make an analogy, today's spectrum use resembles transportation at the beginning of the automotive revolution when we created our highways and interconnection and commerce flowed. The unlicensed bands, used primarily for Wi-Fi, resemble the early shared roadways while the rest of the spectrum system still looks like a series of narrow roads. What PCAST proposes is creating the spectrum equivalent of wide multi-lane superhighways, where the lanes are continuously shared by many cars, trucks, and other vehicles. Spectrum superhighways would be large stretches of spectrum that can be shared by many different types of wireless services, just like cars share a superhighway moving from one lane to another. In contrast to the way we have allocated spectrum, the road system has always let Federal and commercial vehicles share the same highways, with the proviso that government use was allowed to preempt commercial users' rights for reasons of public safety, emergency medical rescue, or national security. There is no reason that the same principles cannot apply to spectrum management. Users of spectrum can accommodate the needs of other users with signals, sensors, and stop lights along the way to ensure there is not a collision with another vehicle. Just as we created the initial transcontinental superhighways in the 20-30 years that followed the 1939 FDR-commissioned blueprint "Toll Roads and Free Roads,"[31] we have the chance to create spectrum superhighways today.

The opportunity exists today to build the first shared-use spectrum superhighways, in the form of broad bands of spectrum, up to 1,000 MHz wide, open to both Federal and non-Federal services. The groundwork for building these spectrum superhighways is laid in the NTIA's October 2011 Interim Progress report, which identifies a number of Federal spectrum bands as being potentially suitable for shared use.[32] The report indicates that the NTIA has already prioritized study of at least six bands below 3700 MHz for possible sharing, ranging in width from 70 MHz (1300-1370 MHz) to 400 MHz (3100-3500 MHz). Most promising would be four bands that total 950 contiguous megahertz between 2700 MHz and 3650 MHz.[33] Combining these bands with the 3650–3700 MHz band that is already allocated (on a low-power and "lightly-licensed" basis) for non-Federal sharing with Federal users yields 1,000 potentially contiguous megahertz, or one gigahertz, of shareable spectrum. Although there will initially be many exclusion zones—like highway lanes under construction in various geographies—making such a wide swath of spectrum available can stimulate needed private sector investment in mass-market technologies, devices and services designed to operate by sharing underutilized capacity in these Federal bands.

Our highest recommendation is that the President issue an Executive Order to prioritize 1,000 MHz of Federal spectrum for review and implementation to create the Nation's first shared-use spectrum superhighways. The recommendations in this PCAST report are intended to provide a foundation for that goal or to institutionalize it as a model start of the next era for spectrum.

Technical assessments of the bands proposed for sharing to effectively evaluate the full scope of the potential risk and benefits are an important tool to ensure that sharing of Federal spectrum with the private sector, previously dedicated solely to the Federal mission, is not harmful to Federal agencies' spectrum-dependent operations.

Although the risks in terms of potential interference with Federal agency mission are clear, the benefits may not be as obvious. Among the benefits that Federal agencies will achieve are increased access to spectrum for training, stimulating the commercial development of spectrum agile technologies. PCAST recommendations will also be creating a stable framework for further system planning without needing to constantly reallocate.

1.2. How Did We Get Here?

Spectrum management has been an integral element of communication policy since the sinking of the Titanic in 1912, which brought recognition of the importance of managing the use of the Radio Frequency (RF) spectrum to ensure reliable emergency, civil, and government communications. At that time, when the rules of spectrum allocation were established, any interference was considered intolerable, and frequencies were allocated to only one service in a given geography. There were a few, large transmitters and many cheap receivers, which had poor ability to screen out signals on nearby frequencies. Interference was not due to the nature of the signals but was the result of the limitations of the receivers. Regulation was designed to suit this model. The primary mechanism to ensure reliable communications was separation of users, with categories of usage defined by assignment of individual frequencies for each user.

The Communications Act of 1934 divided authority over spectrum between the executive branch (Federal Government uses) and the FCC (non-Federal uses).[34] Frequency assignments for Federal spectrum uses were originally administered by the White House Office of Telecommunications Policy, but in 1977 this Presidential authority was delegated to the Commerce Department, specifically to the Assistant Secretary who administers the NTIA. The Assistant Secretary has the sole authority "to assign frequencies to radio stations or classes of radio stations belonging to and operated by the United States, including the authority to amend, modify, and revoke such assignments."[35] The NTIA Organization Act also permits the Secretary of Commerce to allow non-Federal users to access Federal bands by means of an allocation and license issued by the FCC.[36]

The NTIA manages spectrum use by Federal entities with the assistance of the Interdepartment Radio Advisory Committee (IRAC),[37] assigning frequencies to Federal departments and programs for specific uses and specific locations (either as site-based or geographic area authorizations).

Assignments are generally reviewed and renewed on a rolling five-year basis. Although frequency assignments are not permanent, many are in practice open-ended because of the long lifetime of many Federal systems (such as radars and satellites). The NTIA often assigns a number of different agencies and uses to the same frequency band, if they can coexist, and reassigns spectrum from one Federal user to another as programs using various parts of spectrum come and go, according to the priorities set by the executive branch.[38]

According to the NTIA's Office of Spectrum Management, Federal agencies have exclusive use of 18.1% (629 MHz) of the frequencies between 225 and 3700 MHz (traditionally referred to as the "beachfront frequencies"), while non-Federal users have exclusive licenses to 30.4% (1058 MHz). The remaining 51.5% is shared, with Federal use primary and private sector use secondary.[39]

Approximately 80% of the shared allocation—or 40% of the total—have a "dominant" Federal use (e.g., radar, aeronautical telemetry) that under the current coordination regime effectively precludes substantial commercial use of those bands. In other words, nearly 60% of the beachfront frequencies are predominantly allocated to Federal uses, a statistic that illustrates the importance of finding more effective mechanisms to share Federal spectrum.

The FCC is responsible for allocation and assignment of spectrum for all non-Federal uses, including civilian users and state and local public safety uses.[40] Through a public rulemaking process, the FCC first allocates spectrum for general categories of use (e.g., terrestrial broadcasting, fixed or mobile terrestrial communications, satellite services) and establishes "service rules" authorizing use of these bands.

The FCC assigns licenses and other permissions to use bands according to statutory requirements or its own rules. Spectrum is usually assigned to users on a primary basis, though in some cases (e.g., the Amateur Radio Service) secondary operation may be authorized, meaning that the licensed service must accept interference from primary services. The FCC assigns licenses to qualified applicants. Except in the case of exempted categories (e.g., public safety, satellite), the Communications Act requires the FCC to hold an auction whenever there are competing applications for a license. The FCC also authorizes unlicensed uses under Part 15 of its rules, which permits low-powered operations in certain bands subject to noninterference with licensed services. These rules provide the legal basis for Wi-Fi to operate in the 2.4 GHz band, for example.

Congress retains the power to transfer spectrum from Federal to non-Federal use or vice versa. In recent years, several reallocations have directed spectrum from Federal to commercial use, usually through an auction process. The most significant recent reallocation was the Advanced Wireless Services band (AWS-1), in which, following a legislative directive, 45 MHz of Federal spectrum was combined with an equivalent amount of repurposed commercial bandwidth to yield a 90 MHz band that was auctioned for $13.7 billion in 2006.

To clear Federal spectrum for private use, NTIA has traditionally forced all Federal users, usually on a nationwide basis, to relocate from one frequency band to another, although with some amount of sharing during the transition and some permanent exclusion zones to protect immovable Federal operations. Additionally, a number of bands (notably the 900 MHz and 5 GHz radar bands) that are assigned for primary Federal use are shared by the private sector on a low-power, unlicensed basis. As explained earlier in this report, sharing will likely become more common in the future since the Federal bands occupied by legacy users that have been easy to relocate have already been cleared, so that relocation will be very costly in Federal spectrum still held by legacy users.

Presently, NTIA and FCC coordinate interference protections through a bureaucratic process wherever there is a shared band in the table of allocations (and, for that matter, when there are potential interactions among non-shared Federal and commercial allocations). Moreover, engineering analysis supporting interference protection criteria sometimes specifies how Federal uses are to be protected (e.g., through command-and-control restrictions on various operating parameters) rather than what is to be protected (e.g., power density at a protected site).

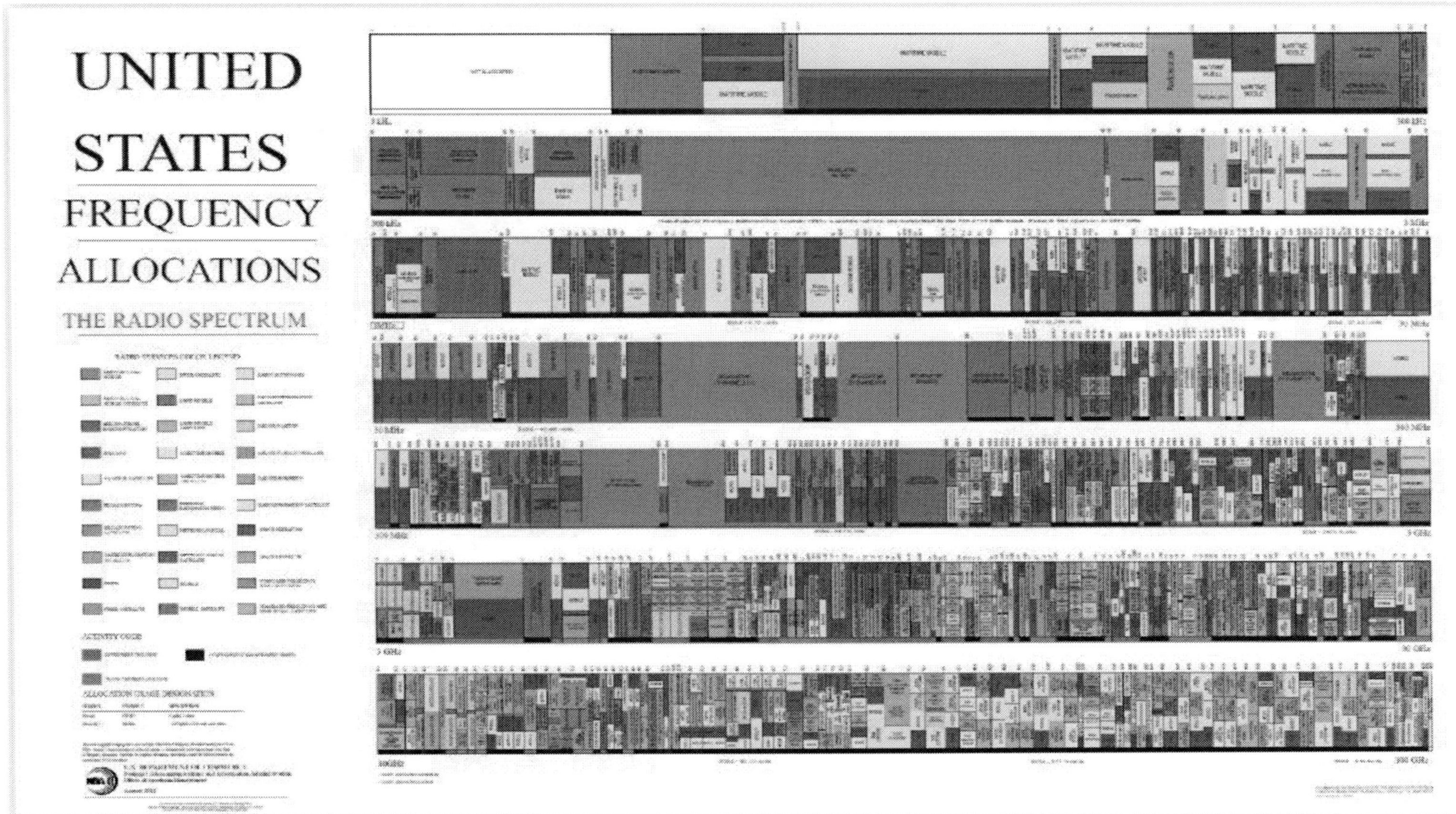

Source: NTIA, www.ntia.doc.gov/files/ntia/publications/spectrum_wall_chart_aug2011.pdf.

Figure 1.3. United States Frequency Allocations in August 2011. This figure shows how spectrum is used, with frequency running horizontally and different colors representing the various applications supported in the bands. Multiple colors in some regions indicate the static sharing of spectrum that occurs today.

This coordination process can be time consuming, highly uncertain, and technically conservative, and can therefore inhibit non-Federal uses of shared or adjacent bands. Unlike the FCC's notice and comment rulemaking process, much of the interagency coordination process, even where national security issues are not implicated, occurs through non-public inter-agency communications. After 100 years of allocations, the current service-specific static spectrum allocation scheme has led to the finely-divided frequency allocation map—the master "zoning plan" of wireless spectrum—exhibited in Figure 1.3. In addition to limiting the amount of contiguous spectrum available for commercial or Federal use, the current regime has created a multiplicity of spectrum borders where underutilized guard bands are imposed to prevent mutual disturbance of services in neighboring bands. In general, the fragmented partitioning of Federal spectrum leads to inefficiency, artificial scarcity, and constraints on current and future Federal and non-Federal uses.

1.3. At the Crossroads, 100 Years Later. What Can We Do Next?

It is time for a change in perspective. Improvements in radio technology mean that "interference" is avoidable or tolerable in most cases today. It has become an excuse for making spectrum scarce. A new, more efficient zoning plan can be made, one in which regulatory process can keep pace with technological change. As this PCAST report argues, calling the problem a shortage of spectrum is a fundamental misunderstanding; rather, the problem is one of how we better manage spectrum. This report calls for a new Federal spectrum architecture and a corresponding shift in the architecture of future radio systems that use it so as to unlock the data-carrying capacity of spectrum in an unprecedented way. The key to the new architecture is to create very wide bands and implement dynamic, real-time, spectrum sharing. The technology to allow multiple users to share the same piece of spectrum, independently and without any advance knowledge of the other users, has become practical in the past ten years. What inhibits greater sharing of spectrum is not lack of technology, but regulatory and economic obstacles. The new architecture we propose does away with those obstacles and opens up Federal spectrum access to many more users. We estimate that in the best circumstances, the amount of effective capacity that can be obtained from a given band of spectrum can be increased thousands of times over current usage through dynamic sharing techniques that make optimal use of frequency, geography, time and certain other physical properties of the specific new radio systems.[41] Optimizing future systems around these concepts, particularly where congestion is likely to be greatest, can yield much more capacity than could even be achieved by clearing and reallocating bands under the current model. Spectrum sharing is beginning to emerge as a mainline approach to spectrum management. For instance, White Space Technology is an example where data systems can co-utilize the television bands.[42] On January 26, 2012, the first commercial deployment of White Space Technology took place in Wilmington, NC.[43] Likewise, radar systems are now able to share with Wi-Fi systems in the 5 GHz band as a result of recent FCC actions that followed a combined effort from industry and Federal agencies, particularly DOD, working alongside NTIA staff, and medical devices (see Figure 1.4)[44] have shared spectrum in parts of the 413–457 MHz range with Federal Government radar and can now share spectrum in the newly designated Medical Body Area Network 2.36-2.39 GHz band with aeronautical telemetry.[45]

Source: Alfred Mann Foundation. aemf.org/our-research/current-focus/neuromuscular-disorders/.

Figure 1.4. Illustrative Medical Spectrum-Sharing Device. This device takes the place of damaged nerves to restore sensation, mobility, and other functions to paralyzed limbs and other parts of the body. It uses smart radio spectrum sharing technologies and can switch frequency bands to operate reliably.

Transitioning to the new spectrum architecture described in this report also implies a transformation of the way that spectrum use yields Federal revenue. This report argues that the United States should shift to a spectrum management model that makes possible a continual stream of revenue instead of onetime auction returns. The revenues would derive from wireless services eager to pay modest fees under a variety of leasing arrangements to obtain spectrum access with varying levels of quality of service and lease lengths, appropriate to their business needs. Chapter IV describes these models in more detail.

1.4. Goals of This PCAST Report

This PCAST Study is entitled "Realizing the Full Potential of Government-Held Spectrum to Spur Economic Growth." It is specifically targeted to looking at spectrum allocated for Federal use, as a logical first step to understand how to move our spectrum policy forward. The question for PCAST consideration in the original statement of task was: "How can advances in situation-aware spectrum-sharing technologies unlock the value of government-held spectrum for commercial use while preserving mission capabilities; and are current Federal efforts in spectrum policy (including R&D investment) optimized to realize this potential?" The ideas and recommendations in this report build on the work inspired by previous government studies of spectrum use and management. In 2002, the FCC formed a Spectrum Policy Task Force (SPTF),[46] which developed a report that recommended a number of practice and policy changes. These included some antecedents to the recommendations in this report concerning spectrum sharing, and interference noise standards, most notably the white space sharing concept.[47] The FCC and industry stakeholders have since advanced White Space from concept to reality for The TV Bands over a period of eight years. In 2010, the FCC issued a crucial order that made white spaces technology practical on a commercial scale.[48] The FCC's National Broadband Plan, issued in 2010, called for expanding opportunities for innovative spectrum access models. These included a recommendation to spur further development and deployment of "opportunistic uses" across more radio spectrum.[49] Subsequently, the FCC initiated a proceeding on dynamic spectrum access technologies, which generated significant discussion among industry stakeholders about new approaches to spectrum management.[50] The FCC's Technological Advisory Council (TAC) and the NTIA's Commerce spectrum Management Advisory Committee Members (CSMAC) has also recently convened working groups on topics such as receiver performance, small cells, data management, and dynamic spectrum access that have informed this PCAST report. In 2011, the National Research Council (NRC) published a report that looked at the key technology considerations and advances in radio technology and offered a number of forward looking policy options around the use of higher frequencies, new approaches to tolerating interference, and designing radios for multiple frequencies.[51] This chapter is organized into seven sections. This section provided an introduction to the problem and a summary of a PCAST recommended solution.

Section II looks at clearing the policy hurdles and specifically talks about the need for a new architecture, a new set of metrics for spectrum use, and the implementation of a Federal Spectrum Access System (SAS).

Section III looks at the technological gains that have been made and need to be made to facilitate spectrum sharing.

Section IV examines how new spectrum and economic models can facilitate new and revolutionary applications.

Section V explains why Federal spectrum is a good place to start with new spectrum management approaches and outlines an incentive program for the Federal agencies to become more efficient in the way they use spectrum.

Section VI describes the testing systems needed to realize the full economic potential of spectrum.

Section VII summarizes the main recommendations and provides an implementation plan and timeline that enables the United States to make an immediate start on evolving its spectrum architecture and management.

Ultimately all countries will have to make the shift to a new architecture because no country has enough spectrum available to meet the exponential increase in demand that comes from the onslaught of these billions of connected intelligent devices, sensors and Internet systems and services. In technology sectors, the first movers capture the bulk of the lifetime profits and those profits fund the development of continuing leadership. The United States is at a crossroads in spectrum; changes in technologies and policies are inevitable. 2012 is a critical year. The Nation must start to change course now to reap the full economic benefits of Federal spectrum.

II. Clearing the Policy Hurdles

As the previous chapter explained, we have created a fragmented partitioning of spectrum that has led to artificial scarcity and constraints on future uses. Because of this history, legacy spectrum assignments remain overly restrictive in view of the ability of today's devices to avoid interference and use spectrum in a more flexible, versatile way.

Taking advantage of these technologies to promote efficient use of spectrum requires policy to evolve toward a new "zoning plan" based on substantial frequency blocks, rather than a multitude of narrow, exclusively assigned bands. A spectrum architecture based on large blocks of spectrum shared among compatible but not necessarily identical services embodies two basic principles embraced by this report: coexistence rather than isolation, and spectrum sharing. Evolution of spectrum management requires a new metric that considers many dynamic variables in the consumption of spectrum capacity and measures spectrum effectiveness, rather than only a static number of megahertz. Once we have adopted a new allocation policy and allow multiple users to share these large spectrum partitions, we will need a real-time access system to oversee and manage spectrum use, in real time, by means of agreed rules for sharing. This chapter describes a new architecture for this allocation system, the metrics that can measure potentially thousand-fold increases in effective spectrum capacity, and the spectrum access system that will govern real time shared use of the spectrum.

Finding 2.1: Sharing of Federal spectrum provides an opportunity to deploy a wholly new approach to Federal spectrum architecture and policy by establishing large shared spectrum blocks, new effectiveness metrics, and coordinated and prioritized Federal and commercial use.

Finding 2.2: Wireless architectures have evolved from a single model of high-power, high altitude base stations to a mix of capabilities, ranging all of the way from base stations to offload onto commercial Wi-Fi. This provides an opportunity to locally exploit Federal spectrum sharing opportunities that would not be otherwise compatible with high power operations (such as LTE).

Recommendation 2.1: The Secretary of Commerce, in collaboration with the Federal Communications Commission (FCC), should establish a mechanism to provide the Federal Government with the ability to manage the sharing of Federal spectrum. Federal spectrum should be divided into substantial frequency blocks with common characteristics, rather than the current narrow band service-specific static allocation scheme. In addition, rather than the current pre-allocation and assignment of spectrum, there should be a new "dynamic sharing" model that makes spectrum sharing by Federal users the norm, and also allows sharing with commercial users. Shared access to Federal spectrum should be governed according to a three-tier hierarchy: Federal primary systems would receive the highest priority and protection from harmful interference; secondary licensees must register deployments and use in a database and may receive some quality of service protections, possibly in exchange for fees; and General Authorized Access users would be allowed opportunistic access to unoccupied spectrum to the extent that no Federal Primary or Secondary Access users are actually using a given frequency band in a specific geographical area or time period. All Federal agencies should be required to cooperate in the implementation of these changes.

Recommendation 2.2: The Secretary of Commerce, working through the National Telecommunications and Information Administration (NTIA) and the FCC, should authorize and implement, directly or through commercial providers, a Federal Spectrum Access System (SAS) to serve as an information and control clearinghouse for the band-by-band registrations and conditions of use that will apply to all users with access to each shared Federal band under its jurisdiction.

The SAS will protect Federal operations from interference while allowing non-Federal users to access underutilized spectrum in Federal bands. Underutilized spectrum capacity in Federal bands should be made available to the greatest possible extent for non-interfering shared use, based on the principle that exclusive assignments should not be taken as a justification for letting unused or underutilized spectrum lie fallow.

2.1. A New Spectrum Architecture

Traditional spectrum management practices by no means maximize spectrum efficiency. Although there is a general perception of spectrum scarcity, most spectrum capacity is not used. An assigned primary user may occupy a band, preventing any other user from gaining access, yet consume only a fraction of the potential spectrum capacity. Indeed, measurements of actual spectrum use show that less than 20 percent of the capacity of the prime spectrum bands (below 3.7 GHz) is in use even in the most congested urban areas.[52]

Unique among natural resources owned by the public, spectrum capacity is infinitely renewable from second to second—that is, any spectrum vacated by one user is immediately available for any other user.[53] The incongruity between concern about a "looming spectrum crisis" and the reality that only a fraction of the Nation's prime spectrum capacity is actually in use suggests the need for a new policy framework to unlock fallow bandwidth in all bands, as long as it can be done without compromising the missions of Federal users and ideally by improving spectrum availability for Federal users. Bringing spectrum management into the 21st century requires starting wherever possible with a clean slate in order to implement a wholly new approach to spectrum architecture and policy. Today, regulators control the characteristics of individual transmitters. In the future, regulators should deal with the spectrum as a system, and will have to address the interaction of elements in that spectrum system, including receivers as well as transmitters. Establishing large, shared spectrum blocks lays the groundwork for policy that views spectrum use as a coherent "ecosystem" rather than an assortment of individual activities. It will also help resolve a technological "chicken and egg" problem, in that there has been a lack of effort to develop low-cost devices that can tune across a wide range of frequencies because there have been no bands in which such devices could be deployed. We believe that such technology will become more available and affordable if spectrum policy formally embraces the provision of wide spectrum bands; in turn, the emergence of appropriate technology will strengthen the case for the new spectrum architecture. In the short-term future, the transition will be eased if requirements to share spectrum in a band are kept within a single octave (or factor of two in frequency). In particular, that will make the technology to retrofit military radios for spectrum sharing by working with multiple bands more affordable and practical.

The move toward higher frequencies and smaller cell sizes is an important development, already under way, that will contribute an essential element of the new architecture. Small cells (see Figures 2.1 and 2.2) of radio coverage using GHz frequencies will increasingly offer a more effective use of spectrum. Higher frequencies are less penetrating, both in air and through building, and are thus suited to services that transmit only over small areas. The use of smaller cells makes it easier to "reuse" a given frequency for geographically separated services, which linearly increases the aggregate bandwidth available to users by increasing the number of access points in a given area. Small cells also fit better with modern devices that are both transmitters and receivers. Smaller cell sizes require more equipment, but with the rapid decline in the cost of wireless devices and the improved availability of high-speed backhaul[54] this is becoming less of an issue.

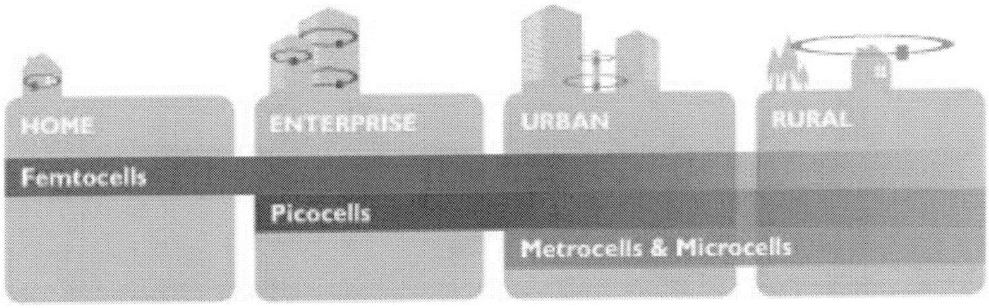

Source: "What is a Small Cell?" Small Cell Forum at www.smallcellforum.org aboutsmallcells-small-cells-what-is-a-small-cell.

Figure 2.1. Small Cells Allow Greater Geographical Coverage. The smallest cells are used in home, enterprise, and urban areas, while metrocells cover larger regions, including rural areas.

The rise of Wi-Fi exemplifies many of the advantages the new architecture can offer. Its tolerance of interference makes it common to see 25 or more independent Wi-Fi networks in a single location, sharing the same spectrum. Although interference has some effect on each network, the effect is minor compared to the massive increase in aggregate throughput. The viability of this approach is demonstrated by the fact that a high percentage of smartphone data traffic is offloaded from cellular networks and delivered through Wi-Fi, even though there are many other devices (besides smartphones) that operate over Wi-Fi networks (see Figure 2.3).

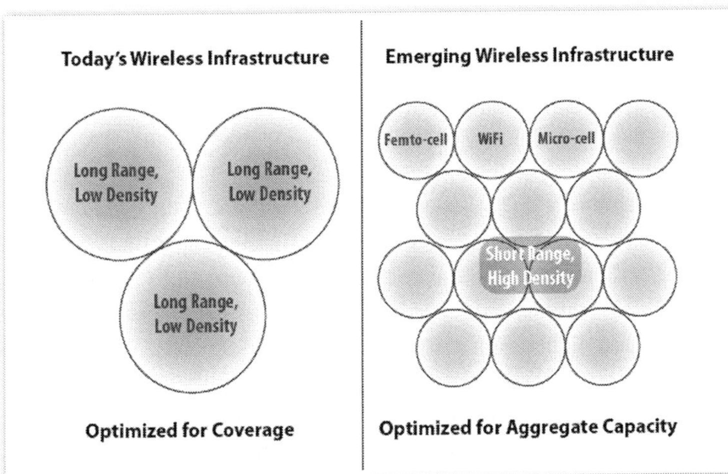

Source: PCAST, adapted from P. F. Marshall, "Scaling, Density, and Decision-Making in Cognitive Wireless Networks." Cambridge University Press (August 2012).

Figure 2.2. Future Architectures will Increasingly Include Dense and Short Range Capabilities to Provide Aggregate Capacity. Decreasing cell sizes results in more cells covering a given area, increasing the network's capacity, or ability to handle a large number of simultaneous users.

Small cell architecture offers an additional motivation for moving to large spectrum blocks. Small cell services typically offer high data transmission rates to their users through the use of relatively high bandwidth; Wi-Fi bandwidth, for example, has grown from 20 MHz to a current draft standard of 160 MHz. Provision of high peak data rates is best accomplished, both technically and from an economic perspective, if there are wide spectrum blocks in which the services can be deployed.

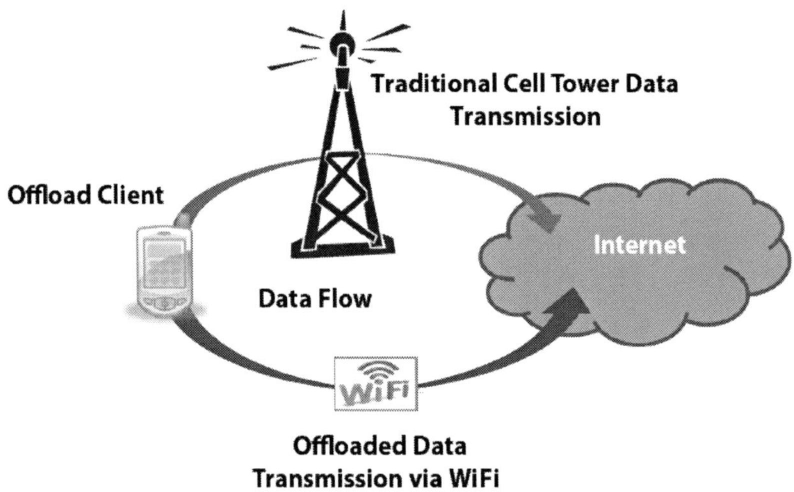

Source: PCAST.

Figure 2.3. Mobile Offload. A large proportion of smartphone traffic now flows through Wi-Fi rather than over the cellular network. Wi-Fi allows for higher capacity because it is a small cell system.

The fact that the higher frequency spectrum used for small cells is unsuited for high power, long range operation has policy implications for assessing the value of spectrum for broadband use. Some spectrum currently excluded from any commercial or non-government use will be viable for shared use by small low power cells in metropolitan areas, even though high power large cell use might still be nationally precluded. The implementation of small cells could make higher frequency spectrum the next "beach-front" spectrum, since wireless infrastructure is now less commonly being "built out" for wide-area coverage but is instead being "in built" for higher aggregate capacity. Small cell technology is also being deployed as an adjunct to traditional carrier infrastructure because of the need to coordinate spectrum use with the established cellular network and to manage the handover of mobile devices from one part of the network to another. In cellular, there has been an evolution of 2G (second generation) to 3G (third generation) to 4G (fourth generation). 1G was analog and first launched in Japan in 1979. 2G was the first digital standard and launched with the GSM protocol in 1991 in Finland. 2G networks allowed data services, offered much more efficient spectrum use, and allowed phone conversations to be digitally encrypted. 3G offered further reliability and speed. Long Term evolution (LTE) can take advantage of advanced topology networks and can create heterogeneous networks (HETNET) with a mix of macrocells with low power nodes. Today, 2G cellular networks cover 85% of the world population and 3G about 35%. We expect this 3G coverage to grow substantially. At the same time, 4G coverage, with LTE as potentially the world's first global mobile phone standard, issmall today, but should increase to about 35% by 2016.[55] Verizon Wireless launched the first large-scale LTE network in North America in 2010.[56] We believe that close coordination with the established cellular infrastructure is not essential to small cell architecture, which will increasingly evolve independently. Voice traffic is becoming just one component of the world of wireless and internet communication, so there is less and less reason for it to be dependent on traditional telephone networks, and the FCC Technology Advisory Council has been analyzing the ways to achieve an orderly transition for the Public Switched Telephone Network (PSTN).[57] In addition, as the number of wireless devices rises, a rising proportion is fixed in place (as in home or business installations) rather than truly mobile, so the problems associated with handover decrease. And as small cell operations become a major service supplier in their own right, rather than a subsidiary to cellular networks, they can increasingly be provided by operators independent of the established cellular infrastructure. (The offloading of smartphone traffic to Wi-Fi, as in Figure 2.3, is a clear sign of this fundamental shift in the way service is supplied). The advantages of wide spectrum bands are evident in the fact that there have always been efforts to keep compatible (similar) services in adjacent bands, in order to reduce or eliminate adjacent channel interference. There is a large body of research on spectrum sharing, and by the end of 2010, there were already 670 Federal spectrum-sharing research projects across 12 agencies (although most of these involve efforts to share spectrum in fairly straightforward geographical ways).[58]

Even so, changes in technology, consumer demand for wireless services, and the entire nature of spectrum use require a much more systematic analysis to determine which services can optimally coexist as "good neighbors" within a large spectrum block. It will also be important to set policy and regulatory principles that promote coexistence among compatible services. A simple method would be to set, for each shareable spectrum band, specific engineering criteria, particularly receiver performance specifications and transmitter emission controls. Existing and potential users of a band would thereby know the conditions under which they are obliged to operate, and could design their receivers appropriately. Similarly they will be able to identify the maximum emission that they can emit in these shared bands. Since users will be in a position to select the spectrum band that is most congenial to their particular services, compatible users will, over time, naturally group together. This system avoids many of the problems that can arise when uses of a band are restricted in a "top-down" way to ensure a degree of compatibility between the users. These traditional controls may lock in spectrum for less advantageous uses, limit innovation, and are not clear and obvious in their interpretation. Even with these controls, new entrants into a band often face litigation and the need for regulatory action. By contrast, determining the eligibility of new sharing entrants solely by their compliance with engineering standards obviates any need to litigate interference issues with all of the incumbent users.

Another approach that will be particularly appropriate for Federal legacy systems that cannot be relocated without great expense and operational disruption would be an analytical system that can classify services according to their operational requirements and thereby select the most appropriate spectrum band for their location, and, given a global view of the receiver and transmitter characteristics of the signals within that band, provide an optimal location for the new signal wanting to access the band.

We are used to thinking about radio spectrum by just the single dimension of frequency, but the criteria that can be used to identify compatible services can include a variety of additional characteristics and parameters. Spectrum can be shared by geography, by time, by economic priority schemes, by code modulation, by polarization, by directionality, etc.[59] These parameters, of course, can be modified by various kinds of behavior, to maximize efficiency, effectiveness, or other objectives, now that radio architecture can be adaptive.

As wireless services are classified by these parameters and behaviors, they should fall into groups or clusters of compatible applications, which can be serviced with similar radio architectures. The characteristics of the service types will partially determine the behavioral parameters. In this vision, these clusters become the "administrative units" of the spectrum.[60]

Ultimately, it is plausible that suitable classifier algorithms can apply these criteria to create compatible service clusters in an automated way. Decisions about which service cluster to place in a given region of "real" spectrum space, i.e. defined by geography, frequency and time, are then a matter of computing values for properly chosen metrics, to achieve the spectrum and national policy objectives. This approach will also need to take into account the costs involved with moving systems.

2.2. The Need for a New Metric for Spectrum Utilization

In order to support a new architecture that promotes spectrum sharing, agile use of disparate bands of reclaimed spectrum, and remarketing of underutilized spectrum, a means to value different blocks of spectrum is needed. A metric that does this must reflect data carrying capacity and some measure of reach or coverage, but these characteristics alone provide an incomplete picture of spectrum use. Metrics are needed that measure spectrum effectiveness rather than simply spectrum efficiency.

More specifically, the metric should measure not only how well a given user can complete a communication, but also the extent to which doing so precludes others from using the spectrum. In other words, the metric should balance the quality of a given use of spectrum with its opportunity cost. Such a metric allows quantitative comparison of differing architectures and protocols for a given block of spectrum, as well as the suitability of different blocks to a given architecture. The metric should be independent of technology as far as possible, and should reward constructive behaviors as well as efficient modulation schemes. The efficiency with which this value is exploited is of great importance, but depends upon what aspect of spectrum efficiency is to be maximized.

Appendix B presents a technical discussion of some possible effectiveness metrics that take into account capacity, communication range, interference range, time, and spectrum precluded to other users. Although there is no established technical criterion for measuring interference tolerance, estimates suggest that the worst case interference range can be reduced by a factor of between 3 and 10. This allows aggregate information rate to rise by 9 to 100 times. Wi-Fi has ad hoc interference avoidance, and commercial 4G technology is already moving in this direction. Much of the increase in 4G capabilities is due to an increased tolerance to interference. As suggested earlier, reducing cell size offers probably the greatest increase in aggregate capacity for most operations. The new metric also demonstrates the impact of receiver protection standards: reducing the interference radius from 1,500 to 150 meters increases the possible aggregate capacity by 100, and a further reduction to femtocell or Wi-Fi ranges of 50 meters provides potentially 400 times as much aggregate capacity. The combination of interference tolerance limits and reduction in cell size therefore has the theoretical potential to multiply available capacity by a factor measured in the thousands or even tens of thousands—far greater than can be achieved simply by adding spectrum, and enough to meet the Nation's broadband needs for years to come. Of course, this is not likely to be fully achieved or necessary in any specific location, but this metric makes clear that the solution to wireless bandwidth is clearly a more complex issue than simply clearing dedicated spectrum. The simplest form of the metric presented in Appendix B has units of *Bits/Area Hertz*, where the area term represents the area precluded to other uses. This measure is appropriate for comparing to the effectiveness of service provided by a a single cellular tower or by Wi-Fi to the same set of nodes, and it indicates why Wi-Fi is so effective despite its limited coverage and availability. Given the fundamental limits of communication theory, the only viable method to achieve density of information transfer is to assure locality of communications. The metric stresses the impact of spectrum reuse, which is fundamental to the development of scalable wireless networks. One of the recommendations of this report is to reverse the historical trend of fragmenting spectrum into ever smaller segments by moving from small spectrum allocations across large areas to larger spectrum allocations in very localized regions. The latter strategy is more effective from both a spectrum and a service perspective. The opportunity to share Federal spectrum provides a policy mechanism to enable U.S. enterprises to implement this scalable architecture, without disruption to current services or Federal users.

2.3. A New Approach for Allocation: A Three-Tier Hierarchy for Access to Federal Spectrum

The technology and governance mechanisms now exist to enable dynamic sharing of underutilized spectrum on a band-by-band basis, while ensuring that primary Federal operations are both protected from interference and able to upgrade their own technologies and use of a band in the future. The system proposed in this report is based on the presumption that all bands with primary Federal users should be open to the greatest practical extent to non-interfering uses. The framework proposed avoids Federal users needing to vacate spectrum and does not preclude expansion of Federal usage. The following descriptions set out our general vision for a three-tiered hierarchy of use (see also Figure 2.4).

"Federal Primary Access" users would register their actual deployments in a database, and in return would be guaranteed protection from harmful interference in their deployed areas, consistent with the terms of their assignment (by NTIA or by FCC). Federal Primary Access users could have exclusive use of the spectrum when and where they deploy networks or systems, but do not have exclusive use where they have not deployed network assets or in locations where, or times when, underutilized capacity can be put to use without causing harmful interference. In other words, Federal Primary Access should be an exclusive right to actual use, but not an exclusive right to preclude use by other Federal or private sector users.

"Secondary Access" users would be issued short-term priority operating rights in a specified geographic area and would be assured of interference protection from opportunistic use (see below); however, they would be required to vacate when a user with Federal Primary Access registers a conflicting deployment in the database. There may be multiple levels of Secondary Access users of spectrum, with different assigned levels of priority, so that some Secondary Access users may be preferred over others, either because of payments (e.g., an auction or user fees), or because of a public interest benefit such as being a Federal user or public safety user.

"General Authorized Access" users would be allowed opportunistic access to unoccupied spectrum if no Federal Primary or Secondary Access users are registered in the database for a given frequency band, specific geographical area, or time period. General Authorized Access users would be obliged to vacate once a conflicting Federal Primary or Secondary Access deployment is registered (or sensed, in the case of bands where devices are authorized to rely on sensing to avoid Federal Primary Access users). General Authorized Access devices should be required to have the capability to operate on multiple bands, using dynamic frequency selection, so that there is no dependency on access to a particular frequency, and so that the device can automatically switch to a different band and not be obsolete if any one band becomes unavailable. Certain bands could also be subject to a device registration requirement, if needed by an incumbent Federal Primary Access system, to facilitate the location and shut down of devices causing harmful interference. The availability of both geolocation databases[61] operating almost in real time along with cognitive radio capabilities (i.e., sensing, dynamic spectrum access), working separately or in combination, make opportunistic access feasible on a band-by-band basis, subject to conditions ("terms of use") that are tailored to avoid harmful interference to licensed operations. The integration of technologies including an automated geolocation database, sensing, signal beacons (which can be used in certain bands to immediately preempt Secondary Access and/or General Authorized Access users) and the band-byband access rules established by the NTIA and FCC, should constitute a comprehensive access system to enable and manage shared access to most Federal bands.

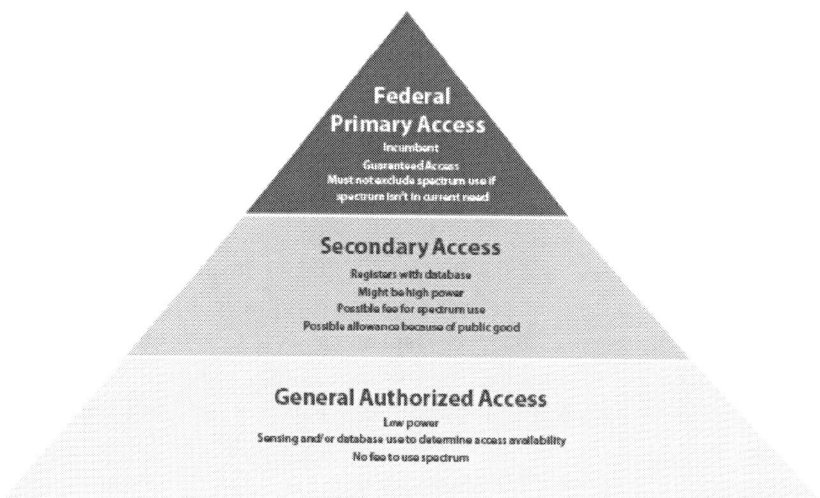

Source: PCAST.

Figure 2.4. Three-Tier Hierarchy for Access to Federal Spectrum. Legacy Federal users have the highest priority, designated Federal Primary Access. Secondary Access is lower in priority. To receive authorization, secondary users must register with a database; they may or may not pay for access, depending on public policy. These users can transmit with high power and have some quality of service provisions. General Authorized Access Users have the lowest prioritization. Depending on policy, they may access the spectrum either by sensing open spectrum or by registering with a database. General Authorized Access allows for only low power transmission, but does not require a fee for use.

2.4. A Spectrum Access System to Govern Shared Access to Federal Bands

We envisage that access to large Federal bands authorized for shared use can be coordinated primarily by registering and communicating with a management database, similar in concept to the White Space Databases certified by the FCC to provide permission to transmit in the TV Bands (see Section 1.3). We therefore recommend that the NTIA should begin immediately to implement a Federal Spectrum Access System (SAS) to serve as an information and control clearinghouse for the band-by-band registrations and conditions of use that will apply to all Federal Primary Access, Secondary Access and General Authorized Access users for each shared Federal band under its jurisdiction (see Figures 2.5 and 2.6). We envision that commercial entities would operate SAS databases accessible to private sector users, as they do for the White Spaces. The NTIA would ensure that the SAS database includes all information necessary to carry out each of several distinct functions useful for facilitating shared access to Federal bands: (1) an inventory of actual operations and required interference protections; (2) band-specific operating rules and enforcement mechanisms to ensure primary systems protection from harmful interference; (3) permissions to transmit; and (4) device authentication.

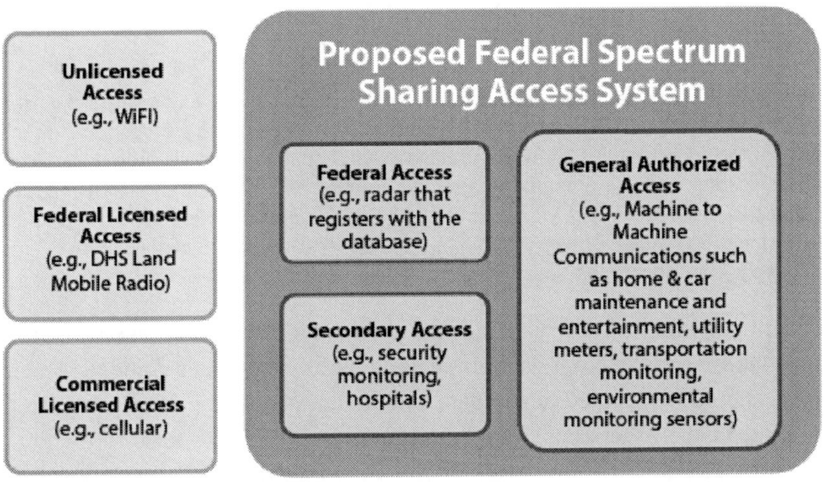

Source: PCAST.

Figure 2.5. Existing and Proposed Additional Federal Spectrum Management Approaches. Today's spectrum users are in three categories: Federal users, licensed commercial users, and unlicensed commercial users. The proposed system will add three new categories. The first is Federal Primary Access, for legacy Federal users that share their spectrum on a first priority basis with other Federal users or commercial users. Conflict is managed by registering spectrum usage in a database. Secondary Access users are Federal or commercial users that have the next priority to shared Federal spectrum. Applications that require higher power and better quality of service than today's unlicensed devices will benefit from this category, although a fee may be required to access this spectrum. The third category, General Authorized Access, has the lowest priority, and supports less critical low power applications such as meter reading or entertainment.

Implementation of the sharing of spectrum is complex and will require extensive efforts to ensure Federal operations are not compromised, including development of interference limits; Federal preemption rights; overhauling Federal spectrum management and frequency assignment process; protections for new Federal systems; and enforcement mechanisms. Development and approval of such methods will include input from National Security Staff. Federal preemptive rights were previously highlighted as a major tenet of this proposal during discussions with PCAST. However, the report does not scope these rights nor address how this concept would be implemented.

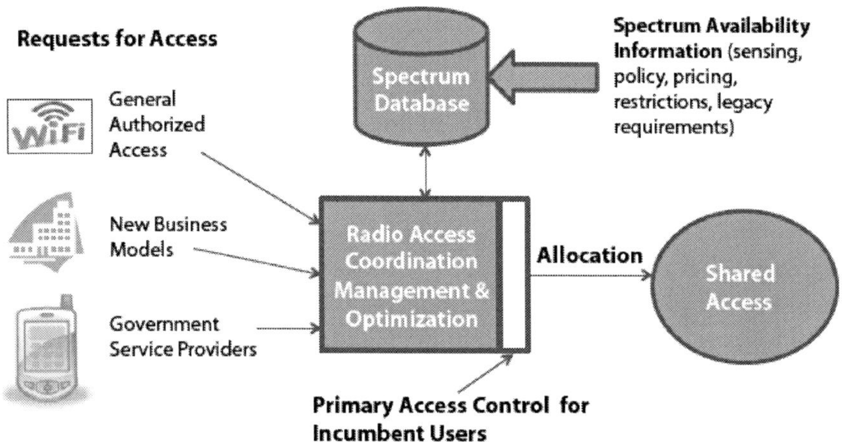

Source: PCAST.

Figure 2.6. PCAST's Proposed Federal Spectrum Access System (SAS). The heart of the proposed SAS is a database that holds information about what spectrum is occupied for a a given location and time; the parameters of the signal, such as power and bandwidth; constraints for specific locations, such as no transmission in blasting zones or along international boarders; and the price for accessing the spectrum. The Radio Access Coordination and Management and Optimization function provides frequency assignments and authorizations. It may work to optimize overall spectrum efficiency over a given region, but above all will insure that legacy Federal retain priority access to spectrum.

Federal Primary and Secondary Access users would affirmatively register operations with the SAS to obtain interference protection. This inventory of primary and secondary uses of Federal bands should be detailed, up-to-date and, as far as possible, transparent to the public. There would need to be exceptions to the transparency requirement for information pertaining to classified uses. Registration data would include the information necessary to determine the availability of a band for shared use, including spectrum actually in use (frequency range), times in use, identity of the user, and as many other operating characteristics as can be safely disclosed. Requiring periodic and automatic communication by devices (or their base station) with the database will ensure that no devices or networks are operating on out-of-date terms of use or without the capacity to be denied access a particular band when necessary. The SAS will thereby enable regulators or other users to verify that the spectrum is being used in a way that is consistent with the terms of use governing access to each band.

While the SAS is agnostic regarding the technology used for radio access coordination, in the case where a cellular carrier is an exclusive secondary to a Federal primary, some technologies will operate more optimally than others. Specifically, cellular technologies that use a single channel for uplink and downlink operation (ie., Time Division Duplexing or TDD) will allow more flexibility than cellular technologies that require a traditional design of paired spectrum where there are two fixed channels with fixed spacing between them (ie., Frequency Division Duplexing or FDD). Since LTE can accommodate both, and LTE chipsets have been created for single channel operation (TD-LTE), use of a single channel will enable the SAS to support continued evolution of wireless carrier capability as we move from voice-centric to data-centric mobile use. Appendix C, Section C.2, expands on this point.

In general, for spectrum sharing between commercial and Federal users to be effective, it will be essential to establish clear, generally applicable "rules of the road" governing the interference environment. In order to protect important Federal missions, NTIA and FCC should revise their coordination process to (1) ensure timely implementation of sharing frameworks, (2) focus on creating good "spectrum fences" rather than dictating "input" parameters, and (3) testing and trusting sharing technology, with appropriate safeguards, to avoid overly restrictive limits on spectrum use. Appropriately revised, the NTIA/FCC coordination process could provide a platform for increased spectrum sharing, rather than a barrier to it.

As described in Appendix C, the SAS includes many features intended to assure Federal agencies operating incumbent primary and secondary systems that shared spectrum access will be governed by band-specific conditions or terms of use that will safeguard their operations against harmful interference, including:

- **Enforcement:** Certain devices could be required to register or to incorporate the capability to receive and switch off a frequency immediately in response to a narrow-band signal beacon (such as during an emergency when a Federal primary or other public safety use requires preemption of a band).
- **Time to Live:** The database should have "time to live" (TTL) entries, similar to the DNS, so that registrations and reservations for use are time-limited and renewed as appropriate. And depending on the needs of the primary users in that band, the TTL mechanism could be extended to allow the database to contact the secondary device and immediately rescind its authorization to use the spectrum.

III. Technology Advancements and Challenges to Solve

Finding 3.1: Spectrum management and regulation is focused on the characteristics of transmission, whereas receiver characteristics increasingly constrain effective and flexible spectrum usage.

Recommendation 3.1: The Secretary of Commerce working through the National Telecommunications and Information Administration (NTIA), in cooperation with the Federal Communications Commission (FCC), should establish methodologies for spectrum management that consider both transmitter and receiver characteristics to enable flexible sharing of spectrum. To safeguard primary Federal users, FCC should require that future non-Federal devices will be permitted to share government spectrum as Secondary Access users only if they are certified to operate within the stated interference limits for the band of interest. Initial specification of protection should be reviewed such that they safeguard new FCC assignments against harmful interference while grandfathering in existing devices and operations.

The previous chapter explained how a new spectrum architecture can enable an evolution away from traditional spectrum management that relied on a combination of exclusive frequency assignment, geographic segregation, and limitations on transmission power to meet the U.S. regulatory standard of avoiding "harmful interference" among spectrum users. However, there is even now no formal definition of what constitutes harmful interference. This lack of rigor in definition, along with a lack of mechanisms to assess the true impact of interference, has resulted in burdensome regulatory proceedings, put the onus on new technologies and service entrants to prove that they are not harmful, and thereby discouraged investment in new products and services. Most interference issues are argued in a very restrictive way, based on analysis of the worst case (from the incumbent perspective) possibility of what might happen. These restrictions are no longer justifiable from a technology perspective. Technology is now available, and will continue to emerge, that can address and manage interference expectations of spectrum users, but spectrum policy and practice have thus far not sufficiently taken advantage of this technology as part of a strategy to create spectrum abundance.

Along with the evolution toward small cell architecture, the emergence of devices that can sense the presence of other transmissions and adjust their operation accordingly has enabled systems to become more tolerant of interference, so that they can function with little or no perceptible loss of quality even in the presence of other signals. Interference tolerance allows systems to coexist more easily in a given spectrum band, and thus increases the efficiency of spectrum use. The huge recent explosion of Wi-Fi services is just one example of this technological evolution. This chapter argues that spectrum management should move away from its traditional focus on control of transmission to a broader approach that includes receiver management, and specifically promotes receiver technologies that are better able to function in the presence of known levels of interference. A flexible approach to receiver management can vastly increase spectrum capacity without endangering the security or functionality of essential systems.

3.1. Technological Advances that Foster "Dynamic Sharing"

The introduction of digital electronics based on transistors and later integrated circuits allowed precision time-based and other forms of sharing.[62] Geographic sharing has also advanced.[63] Such methods for spectrum sharing have been applied, with steadily growing sophistication, to major homogeneous systems, such as the commercial cell phone system and some modern government systems. The challenge and the opportunity now is to apply these techniques more widely, to larger spectrum bands shared by heterogeneous networks—i.e. those that include many different kinds of wireless systems. The ultimate goal is to achieve dynamic heterogeneous spectrum sharing, in which spectrum users can co-exist closely in frequency, time, and geography, dynamically adapt to both the environment and the presence of other users, and do this, moreover, without the massive infrastructure normally associated with a cellular system.

Several recent technological advances make dynamic spectrum sharing practical as a crucial element of the new architecture this report envisions.

Communications Reliability

The purpose of spectrum management has traditionally been stated to be minimization of harmful RF interference between users. A more appropriate objective for today and the future would be to ensure that interference does not cause loss of communications between users. Modern technology has made considerable strides in ensuring that user communications can be accomplished even in the presence of considerable interference. This is already the operational norm for cellular systems in dense urban areas, where many users can conduct conversations and exchange data without any awareness of each other's presence on the system. Technologies that make this kind of coexistence possible include:

- **Link Communications:** Error correcting codes correct for noise effects and random errors. Interleaving and coding techniques can correct errors during bursts of interference.
- **Link Management:** Modern protocols recognize when information is lost across a link, and if necessary retransmit sufficient lost content to recover the corrupted data with no perceptible interruption of service to the user. Furthermore, resources are not wasted on transmission to users with a poor channel or low priority; instead, the system waits for an opportunistic time to transmit and provides better service to those that can receive a useful transmission.
- **Transport Layers:** Transport layers recognize lost blocks of data and request retransmissions from the host computers to ensure that transferred contents are a perfect copy of the original.[64]
- **Adaptive Equalization and Interference Rejection:** Bandwidth-efficient data transmission is made possible by the use of adaptive equalization to compensate for the time dispersion introduced by the channel, and interference rejection techniques can be used at both ends of a link to improve capacity. Adaptive antennas can play a role in interference rejection.
- **Applications:** Many streaming services (both voice and video) are highly tolerant of lost content, and can provide a seamless user experience even in the presence of high loss rates.

These technologies have all emerged since the principles of interference free communications frequencies were established in law, policy, and regulation. They could enable a significant relaxation of the requirements for interference free channels, and would enable massive increases in the density of spectrum use, if policies were adjusted to reflect the capability of these technologies.

Although the recommendations are based on existing, and conservative, use of white spaces databases, there are a number of new technologies that promise to extend the effectiveness of Federal spectrum sharing. These technologies will likely mature and become viable during the implementation of these recommendations, and will enable more dynamic and flexible sharing of spectrum. These technology trends are:

- **Automated spectrum management protocols:** Radios can now be made to recognize spectrum conditions by equipping them with the spectrum management decision tools needed for sharing, including automated interference avoidance and automated, distributed channel selection. Sharing can now be implemented within the device or network in a cost effective, secure, and reliable manner. Early versions of this technology are currently present in the radio resource management of LTE-Advanced and other commercial systems.

- **The general movement to adaptive radios:** Adaptive radios do not fail in the presence of interference or other conditions, but modify their operation to obtain the maximum performance possible. They may presume interference as the norm, rather than the expectation of exclusive, interference-free channels. Wi-Fi and Bluetooth are well-known examples, but this technology should be extended to all Federal and non-Federal devices.
- **Dynamic Spectrum Access (DSA):** DSA systems locate unused spectrum and organize their users to operate within it. DSA systems ensure no interference to other users by scanning and sensing the environment, as the Defense Advanced Research Projects Agency (DARPA) NeXt Generation (XG) spectrum sharing field tests have established. Both the Department of Defense (DOD) and commercial parties have been developing cognitive radio technology (see Figure 3.1) that allows secondary users to operate if they sense that the primary user is not active. For example, the DOD and NTIA collaborated with industry to develop the Dynamic Frequency Selection (DFS) technology that enables 5.8 GHz Wi-Fi devices to detect and avoid military radars.[65]
- **Opportunistic white space radios:** White space radios have a similar objective of making unused spectrum available, but do so through a database of spectrum usage. This technique avoids the need for the sharing device to sense other users, and therefore avoids the possibility that the device might fail to sense some active users, but it does not have the dynamic characteristics of DSA. The FCC is currently validating databases services for sharing unused channel white space.
- **Smart antenna technology:** Multiple-input multiple-output (MIMO) enabled radios have proliferated in the last five years. Placing multiple antennas on a device makes it possible to increase spatial reuse by exploiting multipath effects (signal echoes) for each radio in an operating environment.
  This is especially effective in urban areas where rich multipath scattering can lead to a linear increase in throughput per additional antenna. This technology is in widespread use in Wi-Fi and 4G cellular systems.

The existence of these technologies allows new civil devices that share Federal spectrum to be frequency agile and also suggests that it will be possible to gradually replace legacy equipment operating in the Federal spectrum blocks with agile systems that can embrace sharing. Recognizing the difficulties inherent in this transition, we argue that it also presents a great opportunity because a practical system for spectrum sharing among Federal users will provide significant relief from the growing pressure of spectrum crowding.

Moreover, the inherent complexity of integrated spectrum use through sharing mechanisms provides an additional measure of security and flexibility by allowing Federal users to spread their usage across multiple bands rather than remaining fixed in narrowly defined spectrum allocations.

A promising variation may be to clear and relocate some Federal services and allow others to stay and then share around these services. That variant is consistent with our proposal for a spectrum management architecture that groups compatible systems for most effective sharing.

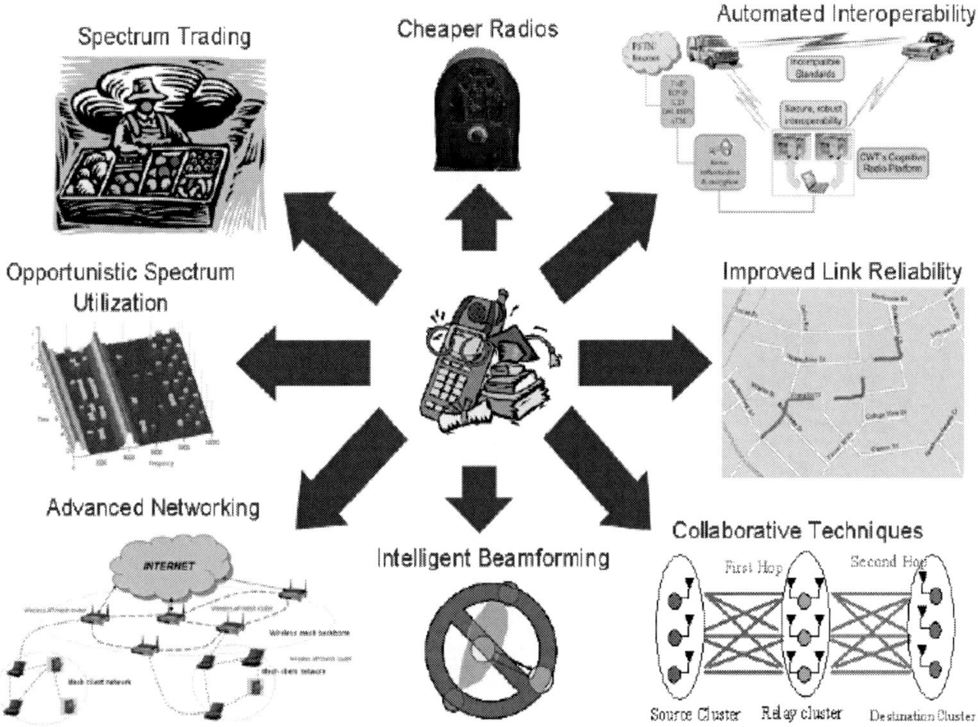

Source: PCAST.

Figure 3.1. Possible Near-Term Cognitive Radio Applications. An abundance of new applications take advantage of new spectrum access techniques using new market concepts; lower cost radios with less need for filtering due to intelligent selection of frequency; higher reliability through radio resource management that uses past information about the reliability of reception at different locations; smart beamforming techniques that permit overlapping cells; collaborative transmission and reception among radios that increase range and reliability; and advanced networking techniques that can better manage unreliable wireless channels.

3.2. The Need for Receiver Regulation

Spectrum management has traditionally focused on the characteristics of transmitters, but receiver performance also limits the utilization of spectrum. Receivers not only receive the signal on the intended frequency but also respond to signals on adjacent frequencies. Inadequate receiver performance can cause signals to "mix," creating false signals on the intended reception frequency or causing the receiver to detune its operation. This is referred to as receiver overload, or desensitization.

Thus the receiver characteristics can limit activities in adjacent spectrum bands (see Figure 3.2).

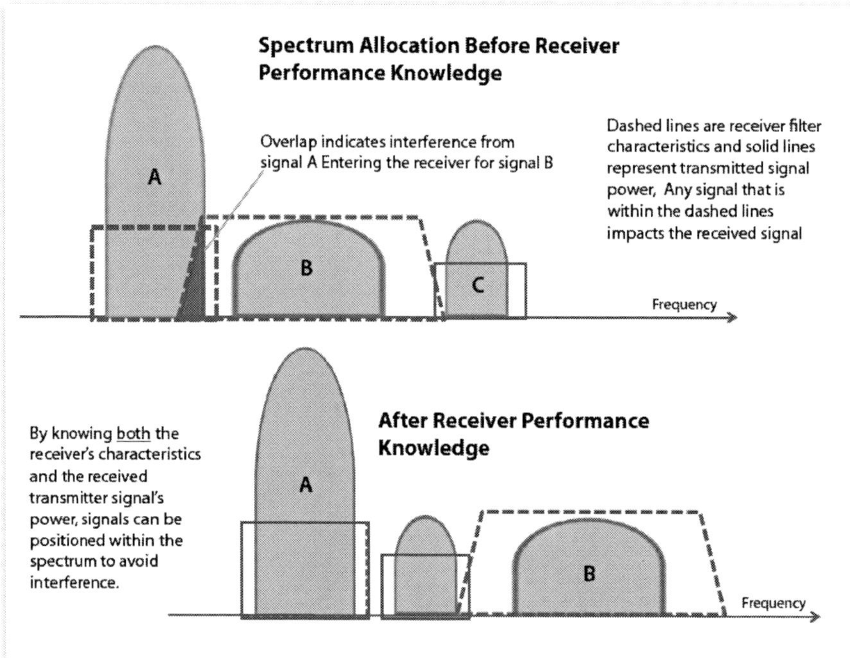

Source: PCAST.

Figure 3.2. Importance of Knowing Receiver Characteristics in Spectrum Management. The upper panel represents three signals in a band; signal A interferes with signal B because it partially lies in the B receiver's bandwidth. A more intelligent choice, shown in the lower panel, allocates frequencies based on the receiver characteristics, transmit power of the signals and received power at each of the nodes, and avoids interference through a different arrangement of the signals in the band.

Recently there have been a number of high profile adjacent signal interference cases that have been major public policy as well as technical issues. The Government had to make major spectrum swaps, allow the purchase of spectrum licenses outside of the auction process, and make major investments in public safety systems due to desensitization of public safety equipment on frequencies adjacent to Nextel's cellular towers.[66] More recently, the adjacent band response of Global Positioning System (GPS) receivers led to an indefinite suspension of LightSquared's plan to utilize a satellite band for terrestrial broadband (see Box 3.1, at the end of this chapter). Interference between garage door openers and military communications systems[67] has left a trust gap between incumbent Federal users and potential non-Federal sharers of Federal spectrum. These are just a few of the higher profile examples. A report of the FCC TAC contains a more comprehensive list of examples in which receiver performance was a significant issue affecting access to spectrum for new services.[68]

In the past, when spectrum was not intensely utilized and there was less need for systems to operate close to each other, adjacent signal receiver-driven interference issues were rare. Looking ahead, we should view the recent LightSquared controversy as an example of the type of conflict that will occur more frequently if receiver-driven interference issues are not addressed. Furthermore, it is essential to address receiver-driven interference issues in order to overcome the trust gap between Federal and non-Federal users and create viable sharing of Federal spectrum.

In order to stimulate innovation and investment in the wireless space, this report recommends against the use of heavy regulation of spectrum and devices to solve receiver-driven interference issues.[69] Instead we propose a receiver management framework that does not mandate additional costs on receivers but provides a framework for defining harmful interference and provides clarity on the requirements that a new entrant must meet to co-exist with legacy systems in adjacent bands. This framework would give device manufacturers freedom to address those requirements as they see fit.

Since the United States, and particularly the FCC, has never engaged in widespread receiver management,[70] we recommend starting with the smallest plausible incremental step: delineating the radio interference that receivers should be expected to tolerate in without being able to make claims of harmful interference.[71] This would define what we call *receiver interference limits* (see Figure 3.3).

A change of regulatory approach that limits the ability of poor receivers to make claims of harmful interference is a necessary complement to existing transmitter regulation. Our goal is not to avoid harmful interference as such, but rather to make it easier to determine which party bears the responsibility for mitigating harmful interference when it occurs. Some might argue that the FCC's current authority to implement any receiver framework, even the lightweight version proposed here, is presently limited under the Communications Act. Moreover, in the recent spectrum legislation, Congress tasked GAO with studying receiver performance and its impact on spectrum efficiency.[72] In light of our recommendations, and recent Congressional interest in receiver performance, Congress may want to consider clarifying the FCC's authority under the Communications Act to develop clear rules of the road for receivers, essentially a receiver management framework.

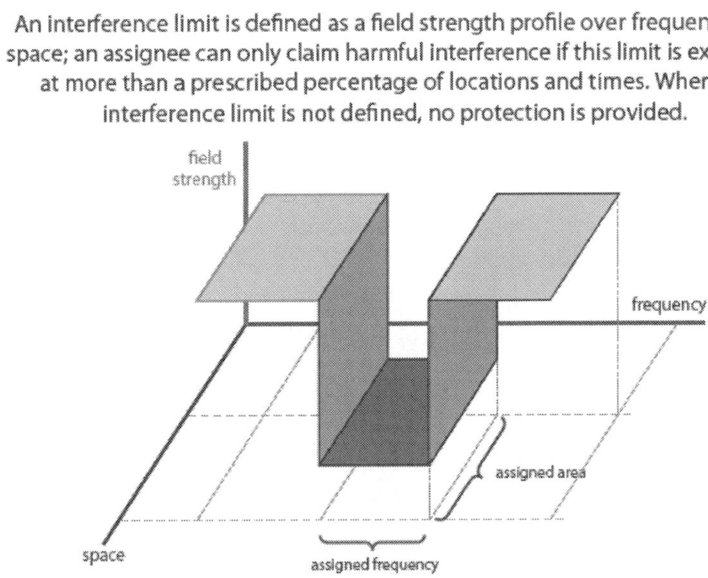

Source: J. P. de Vries.

Figure 3.3. Interference Limits.

Receiver Classes

Different types of receivers require different approaches to receiver management. We have identified five different classes of receivers based on the type of frequency assignment (Federal or non-Federal) and whether the receiver is under control of the spectrum licensee or not. Receivers are divided into three types: licensed operations, decoupled receivers, and Part 15 devices. A decoupled receiver is a device that is not controlled by the spectrum licensee and typically sold to individual users. Examples include television, GPS, FM radio and satellite weather image receivers. For the purposes of this discussion, we will group decoupled receivers and Part 15 devices together since they share key characteristics such as not being subject to a license.

There are then three categories of band occupancy: Federal assignments only, non-Federal assignments sharing with Federal users, and non-Federal assignments sharing with non-Federal users. In the Federal only category, decoupled and part 15 uses do not arise.

In the second category, interference issues are more complicated because of the trust gap discussed above. While garage door openers were technically required to accept interference from military operations in the same band, constituents suffering from the interference were able to create significant political pressure on the DOD to coordinate deployment and activity in the band with the garage door opener manufacturers. While ultimately the cost of modifying, or moving, the garage door openers was borne by the garage door opener manufacturers and consumers, the time and effort involved in managing the political process and frequency coordination was significant. In addition, explicitly coordinating Federal, especially military, frequency use is not possible or desirable in many cases. The concern is that if poor receivers are allowed to proliferate in a band, non-Federal users will acquire the political means to force the DOD to bear the cost of solving any resulting interference problems. For non-Federal users sharing Federal assignments, the FCC must therefore create a receiver management framework that provides the incumbent Federal users with confidence that the new devices will not be able to claim harmful interference from their systems.

Finally we have the third category, which covers sharing of non-Federal assignments by non-Federal users. In this case the FCC must create a receiver management framework that will insure that sharing will not be precluded by poor receivers being deployed. We expect this to be a lightweight framework as market forces will be the primary driver in these cases. Table 3.1 summarizes this classification.

Table 3.1. Three Categories of Band Occupancy and Five Receiver Class

Band Occupancy Category	Receiver Class
Only Federal assignments	All
Only nonFederal (FCC) assignments	Licensed
	Decoupled receivers in licensed service;Part 15 communications devices
NonFederal (FCC) assignments share band with Federal users	Licensed
	Decoupled receivers in licensed service;Part 15 communications devices

Benefits

Setting receiver interference limits that describe harmful interference conditions objectively would enable new entrants to anticipate more accurately the costs of deploying transmitters in a given band. Interference limits would enable the regulator, or coordinator(s) of a shared band, to derive allowable transmission permissions for other operations rapidly and automatically. Similarly, the limits for the receivers of new devices could be easily derived from the permissions of pre-existing transmitters. This would be accomplished without the regulator having to devise performance specifications or impose technology mandates that increase the cost of all receivers regardless of their intended use.

Receiver interference limits would also ensure that bands next to new allocations will be usable in future, even though they might be radio quiet today. Setting the limits high enough that intensive-use allocations such as terrestrial mobile broadband would be allowed in adjacent bands would make it clear that a new licensee should not design their system on the assumption that the neighbors will always be quiet.

In Federal bands, receiver interference limits can be initially set so that existing Federal systems in each band comply with the requirement without any change, thus imposing no cost on existing Federal users. Over time, in order to improve spectrum efficiency and increase spectrum usage, regulators should increase the limits as new devices are introduced and legacy devices are phased out. Appendix D gives a detailed account of one possible implementation of a receiver regulation framework, including progressive steps that can be taken to make sure devices can co-exist together. Section 7.2 sets out an implementation plan for the short term, medium term, and long term horizons.

Box 3.1. The LightSquared–GPS Controversy: An Illustration of the Need for Receiver Management

On February 14, 2012, the FCC, through a public notice, proposed the revocation of its earlier conditional approval of LightSquared's plan to build a satellite-based national LTE network on the grounds that it would cause unacceptable interference to GPS. That decision may have put an end to one long-running controversy, but similar conflicts will inevitably arise more frequently as spectrum use increases and the wireless technologies deployed in different bands evolve rapidly. However, the receiver management strategy recommended in this report can deal with such conflicts in a way that is transparent and equitable for all parties.

The underlying issue in the LightSquared-GPS controversy was that the requirements for building the LightSquared network could not be determined in advance, since there were no objective criteria by which the extent of harmful interference could be predicted. The LightSquared network was designed following FCC rule-makings in 2003, 2004 and 2005 that incorporated extensive input from the public and Federal agencies.

In particular, the FCC adopted recommendations from the GPS Industry Council and NTIA to protect against harmful emissions from Mobile Satellite Service/Ancillary Terrestrial Components operations intruding into other bands, including the GPS frequency bands.

However, the GPS community raised overload interference issues in connection with the FCC's 2011 Conditional Waiver Order, which specified a number of conditions LightSquared would have to meet before offering any commercial service. "Overload interference" occurs when signals from a transmitter disrupt a receiver susceptible to those signals.

In this case, the potentially interfering signals were outside the GPS bands but still detectable by GPS receivers.[73]

The de facto emission requirement for the LightSquared network was therefore set by the level of overload interference that pre-existing GPS receivers could tolerate. Unfortunately, the extent to which GPS devices were expected to withstand interference from the LightSquared system was not specified before LightSquared began to design and build its network.

Thus LightSquared was unable to determine the complete technical specifications—and therefore the true cost—of its network prior to launch.

The receiver management principles we recommend can establish transparency in dealing with such conflicts. If actual GPS receiver capabilities, including susceptibility to overload interference, had been specified in advance, LightSquared could have either designed a network that would not cause interference or else determined that such a design was not feasible or cost effective. Instead, LightSquared discovered the magnitude of the GPS receiver issues only after it had spent billions of dollars.

By defining the interference limit that receivers under FCC and NTIA jurisdiction must cope with and requiring that receiver manufacturers report what interference their devices can tolerate while still delivering reasonable and customary service, the FCC can specify the transmission rights of new spectrum users in a way that would allow them to co-exist with legacy systems in adjacent bands.

This clarity will be essential if we are to avoid numerous recurrences of conflicts like the LightSquared-GPS dispute.

IV. New Application Economy

Finding 4.1: Moving to a dynamic sharing model for Federal spectrum would unlock economic benefits by allowing the private sector to make intensive use of currently underutilized parts of the radio spectrum. A well-designed Federal spectrum policy opens up opportunities for innovation and growth in sectors that are barely imagined, much less well-defined, when the policy choice is made.

Finding 4.2: Sharing of Federal spectrum provides an opportunity to deploy new spectrum management principles such as shorter term licenses that would be appropriate to new and innovative spectrum-based services and products. This provides an opportunity to collect revenue to the Treasury from the private sector for assured use of spectrum.

Recommendation 4.1: PCAST recommends that policies enabling commercial access to Federal spectrum be based primarily on their effects on innovation and growth in wireless devices, services, and associated markets; direct revenue considerations should be treated as secondary. The Office of Management and Budget (OMB) should develop a model to assess future economic growth effects of wireless allocations as well as revenue from increased economic activity. The National Telecommunications and Information Administration (NTIA) and the Federal Communications Commission (FCC) should continue to embrace the current proven unlicensed model and, for licensed spectrum, explore adding new short and medium-term spectrum license models that could both foster growth for these new applications and collect revenue.

A crucial element of the spectrum policy that this report proposes is that it must create the conditions that will foster a vibrant marketplace that can accommodate the widest range of users, from initial venture-funded startups to established service providers. Today's spectrum ecosystem offers most wireless providers the choice between two extremes for spectrum access: shared, unlicensed use with no protection against interference, or exclusive, longer-term licenses with an expectation of automatic renewal.[74] The number of business entities that have the means to participate in costly auctions for nationwide, long term spectrum licenses is limited. Such auctions represented a feasible business model when the capital expense to get started was very large and when the only application was voice services. As today's appetite for unlicensed spectrum use demonstrates, however, adding a variety of "rental" and "lease" options should foster the promotion and validation of highly innovative ideas through short term, low cost access to spectrum (see Figure 4.1). A range of new models could increase not only the number of participants in this market, but also the range of services, particularly data-rich services, that could contribute to economic growth as well as provide Federal revenue through the spectrum rental or leasing mechanism. This aspect of the evolution of spectrum policy will create an opportunity to unleash in wireless space the same innovation engine that made U.S. companies so dominant in the Internet space.

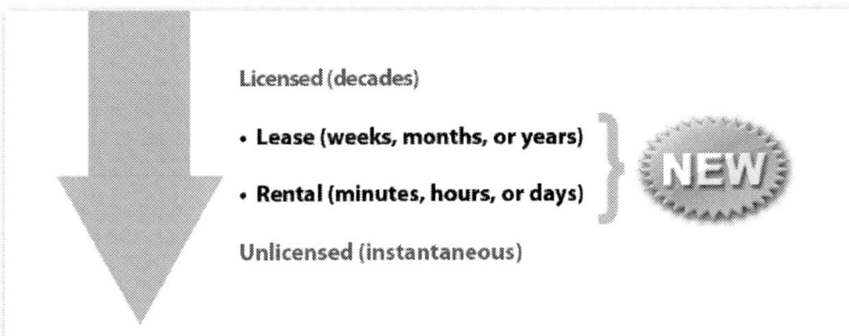

Source: PCAST.

Figure 4.1. Range of Spectrum Acquisition Models. Duration of authorized spectrum access has two extremes today, either decades or momentary. Future spectrum access should provide a wider range of options, including intermediate durations, depending on the application, and quality of service access that is between absolute rights and non-exclusive access.

Spectrum is already a valuable resource, and it will only become more valuable as wireless access becomes more critical to the national society and economy. The new options we propose do not realize this increase in value in terms of a large, one-time payments to the Federal Government, but create the option instead for a recurring revenue stream whose value could increase with time while maintaining the "spectrum liquidity" needed to enable innovative industries to emerge. These options do not carry any specific expectations of either increasing or reducing Federal revenue; rather, they acknowledge that spectrum revenue alone is not the only policy goal. In place of a system that magnifies spectrum scarcity and maximizes the revenue from that scarcity, we envision a regime that creates an abundance of spectrum access opportunities and maximizes the overall value associated with that abundance.

4.1. Evolution of Spectrum Applications

In recent years, many examples of novel spectrum use have emerged. In the ever growing healthcare sector, for example, we see Wi-Fi, integrated with Bluetooth and ZigBee, used to offer an overwhelming array of new services. Many of these applications, such as monitoring patients as they move about a hospital or keeping track of the elderly in their homes, would have been practically science-fiction in the 1980s, when the spectrum they now use was initially opened to unlicensed use.

Similarly, smart grids, and the wireless mesh networks they now use, were practically unknown when the FCC opened up the 900 MHz band for unlicensed use. One recent study of the smart-grid communications market found that advanced meters are being deployed almost three times as fast in the United States as in Europe.[75] Moreover, U.S. smart-grid communications devices are overwhelmingly wireless, whereas in Europe only 20% are wireless. A major underlying cause of these differences is that the U.S. 900 MHz band is larger and has fewer constraints on use and design than its equivalent in Europe. Spectrum policy thus translates into a new market opportunity for U.S. companies, giving them a first-mover advantage in a cutting edge technology market that also supports one of this Nation's most critical infrastructures—the power grid.

By 2020, the connected device market is expected to be dominated not by mobile phones, as it is today, but by machine to machine (M2M) devices—as many as 50 billion of them, by some estimates.[76] Broadly, M2M devices are those that are in constant communication with each other (using wired and wireless networks) without human intervention. Figure 4.2 illustrates the general structure and versatility of M2M networks.

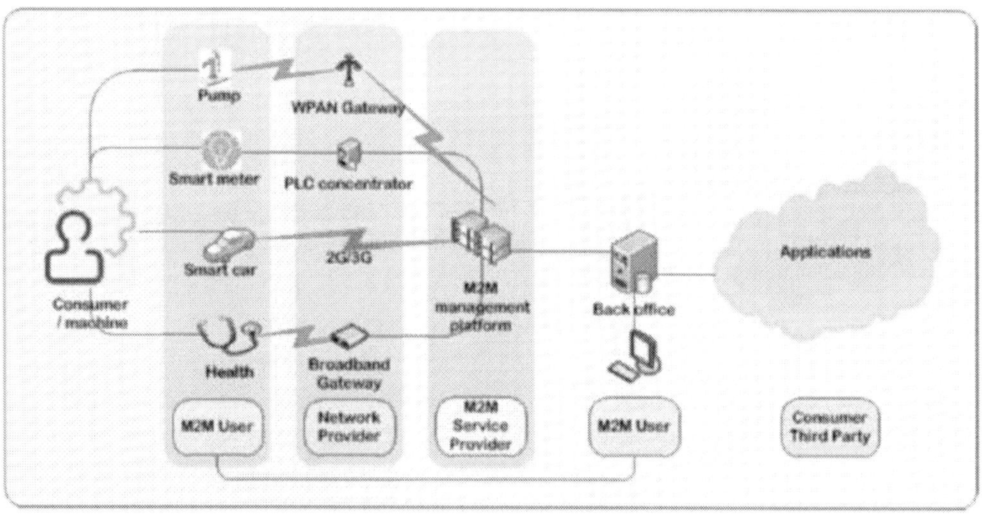

Source: OECD (2012), "Machine-to-Machine Communications: Connecting Billions of Devices", *OECD Digital Economy Papers*, No. 192, OECD Publishing. dx.doi.org/10.1787/5k9gsh2gp043-en.

Figure 4.2. M2M will be supported by a variety and combination of wireless networks, and the overall growth of M2M traffic will grow exponentially.

As mentioned in Section 1.1, a study by Machina Research argues that M2M will generate new revenues and business models as well as enhancing the efficiency and effectiveness of current services, leading to a global business impact worth up to $4.5 trillion.[77] Examples of the economic impact include expected savings of $1 trillion by replacing manual meter readings with smart meters and another $1 trillion from service improvements such as providing clinical remote monitoring for patients with chronic illnesses. The same study lists sixty new potential applications in 13 application sectors. The "top ten" connected applications (Table 4.1) are projected to account for 60 percent of the global business impact in 2020.

Other recent papers[78,79] offer additional examples of novel wireless-dependent markets, including inventory management, access control, mobile payment, and fleet management, as well as broader categories like intra-firm communication or the internet of things. This chapter sets out our thinking on ways to obtain the maximum innovation and economic impact from the wireless sector.

Table 4.1. Top Ten Connected Applications in 2020

Top Ten Connected Applications in 2020	Value to the Connected Life
Connected Car	$600 billion
Clinical Remote Monitoring	$350 billion
Assisted Living	$270 billion
Home and Building Security	$250 billion
PayAsYouDrive Car Insurance	$245 billion
New Business Models for Car Usage	$225 billion
Smart Meters	$105 billion
Traffic Management	$100 billion
Electric Vehicle Charging	$75 billion
Building Automation	$40 billion

Source: GSMA/Machina Research (2012). *The Connected Life: A USD4.5 trillion global impact in 2020. connectedlife.gsma.com/the-connected-life-a-usd4-5-trillion-global-impact-in-2020/.*

4.2. Modes for Private Access to Federal Spectrum

In the past few decades, the two models for giving private users spectrum access—long-term licensed and short-term unlicensed—have been undeniably successful. The first has been central to high-power, large-scale infrastructure applications; growth and innovation have been driven by markets in wireless communications services served by a small number of firms that, for practical purposes, "own" the spectrum license, invest in the infrastructure, and sell services using these assets. The second has low transaction costs, minimal entry barriers, and a reliance on complementary developments in computation and networking technologies and markets; it has spurred the kind of innovation, driven by competition and flexible deployment, described in the previous section.

The FCC has in fact taken significant steps to promote broader access to spectrum by removing regulatory barriers and developing new policies and procedures that have facilitated the development of a secondary spectrum market.[80] More specifically, it has adopted rules permitting licensees that hold "exclusive use" licenses to lease some or all of the spectrum usage rights associated with their licenses to third parties.[81] To motivate the development of innovative applications over the next decade, however, we believe that a greater variety of license model approaches to dynamic sharing of spectrum between users (both Federal and non-Federal) is not only practical, but may be necessary. Many emerging applications may desire or even require predictable quality of service to create effective markets for their products. To create this more versatile market environment, we believe that the NTIA and FCC should develop and experiment with a range of approaches to sharing, certainly including the already proven approach based on unlicensed access but expanding the possibilities to also include new intermediate models of short or medium term prioritized use with the capability of collecting usage fees. Compared to long term exclusive licenses, shorter term secondary licenses have the added benefit of being turned over more quickly, thus motivating a cycle of faster overall industry competition and resulting innovation.

As with larger scale spectrum markets in exclusive licenses, the effectiveness of such intermediate solutions will depend on transaction costs, administrative and regulatory constraints, and the limitations of the sharing rules that are involved. Below we give examples of the kinds of intermediate solutions we have in mind. Since there is no experience with such arrangements, they should be understood not as well-worked out models but as framing models to render the general point more concrete. The particular models developed and tested will, of course, vary with the technical characteristics of the spectrum and the characteristics of the use by the government.

Long-term Licensing would be very similar to current licensing in bands such as those used for personal communications services (PCS) or AWS, where the licensee gets a multi-year (10-15 years) initial assignment. Currently, in the United States, such assignments also have an expectancy of renewal, increasing the value of the initial assignment. Rights for such assignments could be exclusive, or could include well-defined easements for secondary uses, such as low-power unlicensed or pre-emption for public safety use. Long-term licensing has the benefit that companies are assured of spectrum access over a long period of time, facilitating the development of business plans and life-cycle matching with long-lived network infrastructure investments. In addition, licensees would have the incentive and ability to share with other providers provided such sharing does not cause interference either technically or with their business plans.

Medium-term Licensing would operate on shorter timescales (three years or less), on terms negotiated on a case-by-case basis between the government and private users. Those terms would detail the access rights, spectrum quality, and any other appropriate characteristics of the lease.[82] For example, a user might lease the right to use, at a given power level and for one year, a frequency typically used by DOD. Leasing fees could be set by auction, and there would be no automatic right of renewal when the lease expires, although users would be able to compete for renewal by bidding against other potential users. Medium-term leases would give firms the ability to experiment with networks and business models while retaining the chance of extending leases for longer periods. At the same time, the fact that such leases would regularly come up for rebidding might facilitate spectrum access for new businesses and encourage market-based competition among commercial users. Medium-term leases may also provide some assurance to government agencies that if their mission-critical needs change in the future they could have access to spectrum to fulfill their needs, and help build into the spectrum management framework the potential to evolve. Medium-term leases would in general be cheaper than longer term leases.

Short-term Licensing would cover shorter-term spectrum uses, with users agreeing to a fixed set of standards, terms, and conditions to avoid the transaction costs of lease negotiations that may be prohibitive relative to the total value of a shorter term agreement. Quality of access might be variable, and fees would depend on the opportunity cost created by the rented use. The goal would be for markets to determine what "short-term" means, in comparison with more elaborate leasing agreements. For example, a user might rent the right to use spectrum over the course of a few days when the Super Bowl is in a city, or even for the space of a few minutes at rush hour in locations that are dead spots in the user's existing infrastructure. Short-term rentals would allow users to buy access to spectrum with a predetermined quality of service. As spectrally-agile radios become more capable, using short-term access to add temporary capacity for brief periods may not require significant additional capital, so this mode of access could allow maintenance of service during peak demand times. Because the feasibility of the short-term access model may be sensitive to the transaction costs, the SAS should be designed to allow dynamic bidding for brief periods among devices registered in the database.

Because unlicensed use assumes the presence of other devices and no protection, it is generally designed to be robust to interference or intermittent availability. Spectrum use in Federal bands under our new designation of General Authorized Access would, at baseline, be similar to current successful device-centric, rules-based access to unlicensed bands, but users would have to comply by the rules of transmission and provide some mechanism for detection and enforcement of violation of the transmission rules. That is, robustness in the face of spectrum use by Federal Primary Access users could be included in the design of devices permitted to share Federal bands.

As Federal users experiment with versions of more limited, general authorized access use, be they based on databases, spectrum sensing, or some other mechanism, it is important to remember the overall success of simple models that require users to abide by transmission protocols and accept interference from others. Typically, general authorized access would incur no use fee, because power will be low and contention over spectrum small. In some cases, those users may want to move to a secondary priority for a period of time to guarantee quality of service, in which case a fee would be appropriate.

Experience demonstrates that there is significant demand for both long-term renewable licenses and for devices and services based on unlicensed spectrum managed purely through protocols, without the burden of transaction costs. What is unknown at this point is the potential level of demand for short and medium term licenses, or the level of transaction and administration costs that they might bear. These forms of licenses should therefore be tried on a limited, experimental basis to determine when a larger rollout might be beneficial. However, several reasons and sources of insight provide some confidence that such intermediate approaches can be developed and deployed productively; moreover, the availability and flexibility of these economic choices can only help accelerate new innovations and applications, especially M2M applications, that require spectrum:

- Technology is giving devices greater spectrum flexibility. An installation at one frequency can be moved to a different frequency, or even a different band, through low or no cost modifications to the equipment. The loss of a specific spectrum assignment would therefore not equate to the loss of the investment in the infrastructure. Vendors using spectrum in a short or medium term license arrangement can feel fairly well assured that other spectrum will also be available to them when they need it.

- Users will be able to move among the models we explore as their needs or circumstances change. For example, a user could choose to depend on general authorized access operation in some situations, but, if quality of service is important or general authorized access operation becomes too congested, migrate to a protected mode of operation. The same equipment could be operable in both modes, so the ability to change from one kind of spectrum access to another amounts to a form of automatic congestion pricing: a given device can choose to pay a fee for higher priority access only when necessary.
- There appear to be a number of applications that are too local or too small to warrant dedicated spectrum, but which would benefit from or require some form of spectrum access protection. The Utilities Telecom Council (UTC) has made a proposal to use Federal spectrum for use for an electrical smart grid.[83] Responding to growing demand by the medical network community, the FCC recently made spectrum available for low power medical telemetry.[84] These and comparable networks could acquire protected local spectrum sharing rights for critical infrastructure.
- A number of wireless service providers have emerged that provide wireless broadband services in high-density locations, such as airports, coffee shops, etc., using only the unlicensed spectrum. A range of flexible options could enhance this industry's ability to provide a much wider range of products, security, and quality of service options than they can today.
- Over 1,000 commercial Wireless Internet Service Providers (WISPs), as well as hundreds of Rural Local Exchange Carriers (RLECs), currently serve more than three million mostly rural and small-town residential, small business, and public safety customers throughout the country.[85] In large portions of many states WISPs are the only fixed broadband provider. Although WISPs rely primarily on unlicensed spectrum, the Wireless Internet Service Providers Association (WISPA) has advocated for a form of "light licensing" that would give these local operators the added certainty, interference protection and bandwidth they need to provide unserved and under-served areas with Internet service that rivals cable and ADSL (Asymmetric Digital Subscriber Line) offerings in more urbanized areas. While few WISPs have the capital or need for the sort of large license coverage areas sold in FCC auctions, they express strong demand for secondary leases or licenses that could be paid over time from increased subscriber revenue.

This report purposely does not make any attempt to set priorities for the use of the different spectrum access models. They are all examples of possible experiments to be developed and overlaid over the basic argument of this report, that Federal users should make spectrum that they are not actively using available for sharing by non-Federal and commercial users. In some circumstances, a single framework may be sufficient to support Federal and commercial sharing, while in other circumstances we may see all of the models employed simultaneously. The appropriate mix will depend on the technical characteristics (e.g., frequency and adjacency of other uses) of the spectrum at issue, the nature of the government use (e.g., the need for coordination, reliability, and tolerance), the range of possible private uses, and so on. As different agencies, in different bands, experiment with different approaches, it will be important to measure their success in terms of effectiveness, utilization, growth, and value, and disseminate best practices across Federal users.

4.3. Revenue Generation

As this report has emphasized, we do not expect that there will be significant future net revenues from the clearing of Federal spectrum. However, market-based sharing regimes have the potential to generate on-going revenues for the Treasury in the form of spectrum usage fees. The recent history of commercial wireless applications makes any attempt to predict future economic activity highly speculative. The magnitude of any revenues is contingent on how spectrum sharing arrangements evolve, on users' need or desire for quality of service, and on the frictionless design of the revenue collecting mechanisms, and is therefore difficult to forecast. But if we take seriously previously cited estimates of a global business impact in 2020 of up to $4.5 trillion ($2 trillion in savings and $2.5 trillion in products and applications),[86] then even a small penetration of applications paying quality of service fees or other forms of charges could lead to a few billions of dollars in revenue per year. These revenues would help offset the costs of retrofitting and updating various public safety or Federal equipment systems to make them more amenable to dynamic sharing.

The details of how the leasing arrangements described above would produce revenue will have to be worked out as the models are implemented and refined. However, current practices offer some general ideas on how the models might be put into practice. A plausible basic unit for short and medium term leases would be megahertz multiplied by area—i.e. the bandwidth leased for a specified geographic region. A variation on this, already standard in the telecommunications industry, is the "megahertz pop," or bandwidth multiplied by population reached. For example, 20 MHz covering a region containing 500 people represents 10,000 MHz pops. Use of MHz pops recognizes that a very small part of Manhattan is comparable in economic value to a much larger area of rural Kansas. Today, the six major U.S. carriers have long term licenses to over 100 billion MHz pops.[87] We envision that short and medium term leases would give users access to spectrum in much smaller chunks—thousands of MHz pops, say, for short periods of time—and that this flexibility and fluidity would create a vibrant, innovative market with numerous participants.

It is not hard to imagine the kinds of applications that might make use of these revenue-generating models. Under present and near-future architectures, applications that move beyond a fixed local spot and require more-or-less continuous coverage, such as most fleet management solutions and connected car applications, or applications critical in response time, like medical warning devices for acute conditions in mobile patients, or applications beyond the home, such as video downloading outdoors,[88] may require more predictable spectrum access than is offered by present unlicensed architectures, yet not require or have the ability to pay for ongoing or occasional secure access to high-availability spectrum. These novel applications would be good candidates for the new economic models we outlined in the prior section. They would also enormously expand the number of commercial entities able to participate in revenue-generating spectrum activity, compared to the very small number of entities with the means to participate in costly auctions.

Besides new M2M applications in specialized industries experimenting in unlicensed spectrum today, we can also envision that in Federal bands designated with only prioritized secondary use through the SAS, Mobile Network Operators (MNOs) with lower capital expenditures may in the future be motivated to bid on a medium-term license either nationwide or in certain geographical areas.

For the sake of illustration, if we estimate that about $800 billion of the estimated global $2.5 trillion value of products and applications in 2020 comes from U.S. markets for communication and M2M applications, then if even a small percentage of that market makes use of one or more of leasing arrangements described above, reasonable annual revenues will be created. An alternative to leasing as a revenue source would be to have an additional one-time equipment certification fee on new mobile devices, small enough that it would not impact demand. We are purposely not prescribing a specific revenue method as much as illustrating that flexibility can be left to future economists to recommend policy based on future use and supply and demand sensitivity curves. We also choose not to make quantitative estimates of revenue, but we do believe collected revenue will compare equitably or favorably to auction revenue from attempting to clear Federal spectrum, when looking holistically over the next 10 years or longer. If any Federal spectrum could be cleared, we would be forgoing possible small amounts of one-time net fees, but the opportunity to collect a recurring and growing stream of revenue that would grow larger within a few years, would be very compelling. Moreover, these new kinds of revenues could be recognized sooner without similar financial liability, since they do not require many years of band-clearing effort before the bands are accessed by commercial users. Ironically, as significant sharing lowers spectrum scarcity, auction prices may also plummet as an exclusive use spectrum license would not be the only route to entry to offer new mobile services. In that case, the revenue collected from alternative models could easily surpass that collected from exclusive use auctions without imposing much friction on economic growth.

4.4. Societal Value and GDP Growth

The motivation for any revenue-based approach should be to respond to users' for desire qualityof-service, not simply to generate short-term revenue. The guiding principle should be to maximize long-term market growth, but that will be a trade-off for OMB and National Economic Council (NEC) study and analysis. A recent detailed study conducted for the European Commission by SCF Associates viewed "new" spectrum, available by enhancing sharing, as the equivalent to an ongoing new economic stimulus, and estimated that implementing spectrum sharing over 400 MHz of spectrum would be equivalent to injecting 800 billion Euros of economic stimulus into the European Union (EU) economy. Their study also concluded the estimated resulting value to the EU economy between 2012 and 2020 to be at 888 billion Euros with a margin of error of +/- 50%.[89] Like our report, their study asserted the following:

- Sharing networks would utilize a lighter form of alternative infrastructure compared to traditional cellular;
- The opportunity to provide sharing technology would foster innovative startups offering services in shared geographic clusters, new startups fostering cognitive radio technologies, and new database systems similar to our SAS;

- New consumer electronics and the M2M market would quickly move into the new sharable spectrum and start contributing to economic recovery after 2017;
- End user devices would start to benefit from flexible RF capabilities, providing markets for these devices and an unique opportunity for American technology innovation. As with all consumer electronics, a mass market for these devices would itself lead to development of lower cost, high volume technologies, which would further benefit from this flexible regime;
- Federal, public service, and broadcast spectrum would all be part of a primary licensed access system with shared spectrum;
- As spectrum scarcity recedes with significant sharing, auction prices would drive significantly lower as an exclusive use long term spectrum license is only one of many ways to enter the market;
- The indirect effects on the economy would be significant, driving charges to consumers and businesses lower, resulting in even larger increases in mobile usage, and significant benefits of economic efficiency.

The study also found that availability of shared spectrum would have significant and broad societal benefits and open up a wide range of applications, especially improving services for education, health, and social support.

V. STARTING WITH FEDERAL SPECTRUM

Finding 5.1: There is no incentive system today for Federal Government agencies to be efficient in their use of spectrum or to share spectrum allocated to them with the non-Federal sector.

Finding 5.2: A public private partnership (PPP) is the best mechanism to ensure that optimal use is made of the Federally-held spectrum and of related investments in spectrum research and testing.

Finding 5.3: International harmonization of spectrum policies is essential to product innovation, interoperability and roaming, spectrum efficiency, and cross-border frequency coordination.

Recommendation 5.1: PCAST recommends that the White House Chief Technology Officer (CTO) with senior officials at an equivalent level from the National Security Staff (NSS), the Office of Management and Budget (OMB), and the National Economic Council (NEC) formalize a Spectrum Management Team (SMT) to work with the National Telecommunications and Information Administration (NTIA), the Federal Communications Commission (FCC), and the major Federal agencies that use spectrum to carry out the President's directive.

Recommendation 5.2: PCAST recommends that the NTIA, working with the SMT and Federal agencies, reexamine the partitioning of Federal spectrum usage in light of current and emerging technology. One objective of this reexamination is to aggregate current spectrum partitions to create substantial frequency blocks in order to facilitate sharing through common technical use rules.

Recommendation 5.3: PCAST recommends that the President indicate that all Federal agencies should cooperate with the SMT and NTIA to establish and implement a government-wide process and mechanism to share Federally-held spectrum. Within one year, the SMT working with the NTIA should formulate concrete 5-year and 10-year goals for Federal spectrum sharing opportunities in order to recommend to the President how to appropriately update his 2010 goal of making 500 MHz of Federal and non-Federal spectrum available over the next 10 years.

Recommendation 5.4: PCAST recommends that OMB, working with the SMT and NTIA, take steps to implement a mechanism that will give Federal agencies incentives to share spectrum. Such a mechanism would accurately internalize the opportunity cost of Federal spectrum resources and manage them over long time horizons using a "currency-like" accounting, allocation, and incentive system ("Spectrum Currency").

Recommendation 5.5: PCAST recommends that OMB should implement a sustainable funding mechanism to foster a Federal spectrum sharing system. The existing Spectrum Relocation Fund should be redefined as a revolving "Spectrum Efficiency Fund" that recycles private sector payments for use of Federal spectrum into reimbursements to Federal agencies for investments that facilitate spectrum sharing and enhance spectrum efficiency. Congress should allow the Fund to reimburse qualifying costs by any Federal service, not just those in revenue-generating bands.

Recommendation 5.6: PCAST recommends that the President appoint an advisory committee of industry executives (e.g. CEOs), to be known as the Spectrum Sharing Partnership Steering Committee (SSP), to advise the SMT on a policy framework to maximize commercial success, centered on a public private partnership for sharing Federally-held spectrum, and implementation milestones that lay the groundwork for the first spectrum superhighways.

Recommendation 5.7: The United States, represented by the Department of State with advice from NTIA and the FCC, should make international harmonization of spectrum allocations to wireless broadband, particularly in bands used or planned to be used for mobile broadband applications in the United States, a key element of the U.S. position at the 2015 World Radiocommunication Conference (WRC-15) and in bilateral and regional discussions with its own neighbors, Mexico and Canada.

Federal spectrum provides a unique opportunity for the United States to demonstrate how a new model for spectrum management can transform scarcity into abundance, and moreover to establish a leading position in the technology that will ultimately be deployed worldwide as all countries of the world are forced to adopt more aggressive spectrum sharing as a necessary policy.

Sharing Federal spectrum will actually provide more spectrum to Federal users than the current process of spectrum clearing and auction and will therefore provide a sound basis for continued operation and future expansion of Federal systems nationwide. Under the auctioning system, Federal agencies typically have no opportunity to deploy new technology or systems, and their usage is essentially constrained to what was in existence at the time the allocation decision was made. Sharing will also provide the flexible, predictable, and timely spectrum access that will allow the Federal sector to develop and adopt innovative technology and services. Finally, as we have described earlier, opening up Federal spectrum access to commercial users will stimulate economic growth and innovation.

Carrying out the President's Agenda will require the reinvolvement of the White House in spectrum policy, an incentive system to motivate the Federal agencies to share, and a funding mechanism to retrofit and improve the effectiveness of their equipment and reward agencies which are early adopters.

5.1. Prioritizing the 1,000 MHz Spectrum Superhighway

The cornerstone of the new spectrum architecture is the creation of wide bands of spectrum. As stated in Section 1.1, we strongly recommend that spectrum from 2700 to 3700 MHz be prioritized as the basis for the Nation's first spectrum superhighway.

The NTIA's October 2011 Interim Progress Report indicates that the NTIA has already prioritized for consideration for shared use at least six bands below 3700 MHz, ranging in width from 70 MHz (1300-1370 MHz) to 400 MHz (3100-3500 MHz) (see Tables 5.1 and 5.2).[90] Most promising would be the combination of the four bands identified in the Interim Report that total 950 contiguous MHz between 2700 MHz and 3650 MHz. The Interim Report also lists 3550–3650 MHz as a potential band for "licensed non-federal exclusive use." However, as mentioned in Section 1.1, the agency's 2010 Fast Track analysis indicated that if this band were auctioned for high-power, wide-area use consistent with current commercial wireless business models, non-Federal use of frequencies from 3550 to 3650 MHz would be excluded in an area roughly 200 miles inland around the entire coastline of the United States. Dedicating the 3550-3650 MHz band to small cell, low power use could allow for significant reduction or even elimination of the exclusion zones. Indeed, the Chairman of the FCC recently proposed that this band be dedicated for small cell use.[91] The band from 3650 to 3700 MHz is already allocated for non-Federal sharing with Federal primaries on a low-power and "lightly-licensed" basis.

Putting all these bands together would therefore yield a potentially contiguous band, 1,000 MHz (1 Gigahertz) in extent, suitable for shared, opportunistic access conducive to at least very low-power, small cell operations (see Table 5.1).

The bands we propose have characteristics that make them favorable as the early venues for sharing, as recent trends illustrate. Current Wi-Fi services have become highly congested in the 2.4 GHz band. The 5.8 GHz band, although less congested, has significantly less range and object penetration, and only a decade ago, low cost technology for such high frequencies was unimaginable. Once 5.8 GHz became available, however, new technology and products that can exploit it quickly emerged, so that 5.8 GHz has become highly popular for Wi-Fi. The 2.7 to 3.6 GHz band should be even better: it offers significantly greater propagation than 5.8 GHz, resolves the congestion issue (at least initially), and in many areas may be suitable for higher power operation than could be permitted in the Wi-Fi band. A reduction of frequency to 2.7 GHz represents the equivalent of a 460% increase in power, and would double the range of an identically operated Wi-Fi device. The lack of available product in many of the bands under consideration for spectrum in this study is the very incentive for disruptive innovation that this new spectrum policy is intended to foster. The opportunity provided to innovators is not to create just another Wi-Fi band, but to develop services and products that could not be supported within the limits of the current Wi-Fi bands, may not be tolerant of interference, or may require assurance of spectrum exclusivity.

Table 5.1. Federal and Shared Bands Under Investigation for Shared Use

Federal and Shared Spectrum Bands Under Investigation		
Frequency Band (MHz)	Amount (megahertz)	Current allocation/usage (federal, non-federal, shared)
406.1-420**	13.9	Federal
1300-1390**	90	Federal
1675-1710*	35	Federal/non-Federal shared
1755-1780*	25	Federal
1780-1850	70	Federal
2200-2290	90	Federal
2700-2900**	200	Federal
2900-3100	200	Federal/non-Federal shared
3100-3500	400	Federal/non-Federal shared
3500-3650*	150	Federal
4200-4400** [4200-4220 & 4380-4400]*	200	Federal/non-Federal shared Federal/non-Federal shared
Total	1,473.9	

*Bands selected for Fast-Track Evaluation. For purposes of future analysis, 1755-1850 MHz—consisting of 1755-1780 MHz and 1780-1850 MHz—will be assessed as a single block.

**Band obligated by U.S.-Canada or U.S.-Mexico bilateral agreement(s).

Source: NTIA (2011). *Second Interim Progress Report on the Ten-Year Plan and Timetable.* Table 2-1, p. 4.

The suggested priority of sharing within a 1 GHz range of frequencies will enable rapid application of this spectrum. To be effective, devices operating in this sharing regime will have to be flexible enough to move their services in response to local and temporal Federal usage patterns. The technology to operate across a wide range of frequencies is present in most military communications equipment, but is not generally provided by civil equipment, which is typically manufactured only for the exact regulatory band in which it will operate. The focus on less than one octave of frequency coverage will minimize the cost of manufacturing equipment for this band. Although this equipment will initially be more expensive than conventional "fixed" band devices, the savings of billions of spectrum cost will enable a new economics in communications, and enable services that are less cost sensitive to spectrum access costs.

As mentioned earlier, a number of industries, such as medical instrumentation networks, transportation, and smart grid connectivity, have sought dedicated spectrum. Other applications, such as M2M communications, are not well suited to the commercial service offerings that are currently available, and are highly influenced by the characteristics of current spectrum policy. The FCC has wisely avoided further fragmentation of the spectrum by dedicating spectrum solely to these specific uses, since there are likely many such applications and all cannot be accommodated in dedicated spectrum. However, the proposed sharing regime will enable these services to immediately share Federal spectrum, and if needed, rent dedicated spectrum to support their missions, without the necessity to obtain regulatory action, or preclude other uses of the spectrum.

Typically, bands in the range of 100 MHz to 1 GHz have been considered to have the highest economic value because they have high propagation range, building penetration, weather and atmospheric performance, and power efficiency. However, bands in the range of 1 to 5 GHz, although they offer lower performance on those parameters, offer better bit capacity for data rich applications, not only because the channels can be wider, but because they offer better reuse. They also require smaller antennas to match the more local nature of the applications. Ultimately, therefore, the Nation would benefit from having the capability for sharing in a mix of Federal bands at all frequencies.

5.2. Evolving the Federal Spectrum Management Organization for the 21st Century: Re-involving the White House

The Executive Order of 2010 to find 500 MHz of spectrum and the expectation of further spectrum needs in the future confront NTIA leadership with a significant challenge to match its level of responsibility to execute Presidential mandates and its level of authority in dealing with the 65 Federal departments and agencies that use spectrum and 19 Federal agencies on the IRAC.[92] This mismatch will only be exacerbated in a world where spectrum sharing becomes the norm.

Our recommendation for greater White House involvement in spectrum policy and managements represents a partial return to the structure that existed before the creation of the NTIA in 1978. At that time, the White House Office of Telecommunications Policy was responsible for telecommunications policy making and spectrum management. Support for spectrum management, which included frequency allocations and assignments, came from the Commerce Department's Office of Telecommunications.

Various groups in the White House have a stake in the President's agenda regarding spectrum: OMB in allocating funds for spectrum efficiency improvements; the Office of Science and Technology Policy (OSTP) on general spectrum policy; the National Security Staff (NSS) regarding spectrum's role in the maintenance and improvement of national security; and the NEC on the importance of spectrum for innovation and economic growth. PCAST proposes that these four groups come together to formalize a White House Spectrum Management Team (SMT) that would work with the NTIA Administrator to bolster NTIA authority and execute the President's agenda. We recommend that the White House Chief Technology Officer (CTO) take a leadership role, working in concert with the Deputy CTO for Telecommunications, and that representation from the NSS, the OMB, and the NEC be at a similarly senior level. Within one year, the SMT working with the NTIA should formulate concrete 5-year and 10-year goals for Federal spectrum sharing opportunities in order to recommend to the President how to appropriately update his 2010 goal of making 500 MHz of Federal and non-Federal spectrum available over the next 10 years.

We also recommend an evolution of the role of the Policy and Plans Steering Group (PPSG), an interagency organization convened by NTIA in response to the President's November 30, 2004 Executive Memorandum directing the heads of executive agencies to implement the recommendations of the Spectrum Policy Initiative.[93] To form a truly effective steering group, we believe the National Science and Technology Council (NSTC)[94] should add a new (sixth) subcommittee on Spectrum, co-chaired by the NTIA Administrator and the Deputy CTO for Spectrum. The PPSG should then take direction as a resource to the NSTC.

The IRAC should remain as is and continue to work in concert with the technical experts at the NTIA.

Requiring the Agencies to Share Data More Effectively with the NTIA

The NTIA is responsible for assigning radio and radar licenses for Federal users. However, responsibility for verifying the need for radio assignments and ensuring the accuracy of data describing assigned spectrum uses lies with the Federal agencies themselves.

In April 2011, the Government Accountability Office (GAO) issued a Spectrum Management Report[95] that found that "NTIA's data management system is antiquated and lacks internal controls to ensure the accuracy of agency-reported data, making it unclear if decisions about Federal spectrum use are based on reliable data." As a result of the GAO report, CSMAC[96] and the IRAC Frequency Assignment Subcommittee (FAS) are providing recommendations to improve and strengthen NTIA's spectrum data management process. We believe the CSMAC and FAS recommendations relate well to our recommendations concerning reorganization of Federal spectrum management. Specifically, accountability and enforcement of agency spectrum data collection assignments for the NTIA should be handled through the proposed White House SMT, and certification of the quality of the data should come from joint signatures of the newly proposed NSTC spectrum subcommittee and the IRAC technical representatives.

Creating a Robust Framework for Dispute Resolution

The Middle Class Tax Relief and Job Creation Act of 2012[97] created a framework for disputes associated with clearing and reallocation from Federal to commercial use, but in the future, as spectrum sharing (across frequencies, time, and space) expands, the need for a clear and more encompassing dispute resolution framework will loom much larger. The need for such a framework is important both to Federal and civil users as they invest in strategic planning for future spectrum usage models that include significant amounts of sharing. Conversely, the lack of such a framework will likely prove an impediment to a growth in sharing. An absence of disputes would suggest either that there is too little sharing activity or that the mechanism for resolving disputes was inefficient and so expensive, unpredictable, or time-consuming that parties opt not to engage in it. Options may include a variety of future alternate dispute resolution procedures, including creation of a spectrum court modeled on administrative proceeding used by the General Services Administration (GSA). In the interim, the SMT can serve as the Federal adjudicator of spectrum sharing disputes when the Federal user is primary, but certainly these rules would have to be decided in advance through negotiation with the FCC. Appendix E offers some further ideas.

5.3. Incentivize the Agencies to Use Federal Spectrum More Effectively

The vast majority of inputs (such as employees, automobiles, and gasoline) that a Federal agency utilizes to produce services are procured through a market process. That is, the agency is given a budget and can search for the best values consistent with delivery of the required service. Such a process applies to the plant and equipment (e.g., antenna towers and radio transceivers) that an agency needs to make use of spectrum.

However, no such incentives exist in the way Federal users obtain access to spectrum itself. Under the current "command and control" system, Federal users obtain no reward for reducing their own need for spectrum, for sharing spectrum with other agencies, or for sharing spectrum use rights with non-Federal users even when such sharing would be socially optimal. In addition, the absence of pricing signals that would push agencies toward making capital investments to improve efficiency over time tends to build up larger problems in the future: agencies have little or no reason to invest in technologies that could improve spectrum efficiency because they see little or no benefit from any resulting economies. Moreover, uncertainty over their future need for spectrum, coupled with the lack of smoothly functioning market for Federal spectrum, creates an incentive for agencies to hold on to whatever spectrum they already have: An agency that releases spectrum freed up through efficiency improvements might later find itself short of spectrum as new demands arise, yet unable to get the spectrum rights back. Lastly, the lack of spectrum pricing means that no visible budget expense is associated with overall Federal spectrum use, and thus hides the true social cost of that use, which is measured in terms of other uses of the spectrum that are precluded by current Federal use (the "opportunity cost").

Requiring Federal agencies to purchase spectrum rights through a market mechanism would go a long way toward achieving transparency, accountability, and efficiency in Federal spectrum use. It would therefore be desirable to move quickly to a market mechanism so that Federal uses reflect their true social resource cost. There is, however, a long history of failed attempts to implement significant reforms in Federal spectrum use. As we explain below, a system of spectrum fees is in principle feasible, but likely to run into practical difficulties that would render it ineffective. Instead, we propose a model relying on an artificial currency ("spectrum currency") that would allow agencies to participate in a spectrum market within the Federal Government.

Spectrum use fees would be monetary charges levied on agencies for spectrum use and paid to the U.S. Treasury. Use fees would be similar to rent paid to the GSA for office space in government-owned buildings. The length and scope of such rental agreements vary widely. Similarly, spectrum usage fees could vary by level of priority (e.g., primary or secondary use), the length of time the user was utilizing spectrum rights, the geographic scope of the spectrum rights, among other dimensions.[98]

However, the introduction of spectrum fees would not necessarily remove or even significantly diminish the obstacles individual agencies face in trying to evolve their spectrum use in ways that would maximize efficiency by the Federal Government as a whole. In particular, an agency would legitimately fear that if it were to relinquish $500 million of spectrum use, and reduce its fee payment accordingly, it would later see its budget reduced by much of that $500 million and therefore see little or no benefit for its efforts. For that reason, we do not think a spectrum fee system is likely to be an effective way to promote Federal efficiency in spectrum use.

"Spectrum Currency" is our name for a synthetic currency that would be an alternative to spectrum fees that we believe could provide a positive incentive for change. Spectrum Currency would act as an accounting, allocation, and incentive system that governs Federal agencies' use of spectrum and attempts to motivate them to adapt their systems to operate in large allocation superhighways with dynamic sharing of other Federal and commercial systems.

There are several related problems that Spectrum Currency could help solve. First, it provides a way of baselining the relative spectrum "consumption" of any one agency and application in comparison to another. Second, because it is a synthetic currency, it provides a much longer-term horizon than the annual budget cycle for the NTIA and agencies to properly plan to shift spectrum allocations between agencies and applications, and through sharing or give-backs, to the private sector. Third, by creating an incentive system, it encourages government spectrum users to both retrofit their current systems to allow sharing in the near-term and ultimately to replace or move their systems to operate in the new architecture. Fourth, by applying new metrics for spectrum effectiveness (Section 2.2), it starts to let us calibrate how spectrum is currently used in the public sector in contrast to the private sector, especially to the extent that they preclude others from using the spectrum when it is available; it therefore offers a truer value to Federal spectrum. Once a Spectrum Currency baseline is established, it could be used in NTIA and Federal agency planning cycles to focus on ways to improve their effective spectrum use by lowering their future Spectrum Currency needs. We would recommend that the OMB reward early adopter agencies by developing a mechanism that allows agencies to trade Spectrum Currency for the actual dollars that will help them evolve their systems so as to lower their Spectrum Currency requirements in the future. We believe the Spectrum Currency system can promote improved spectrum effectiveness in a direct and easily managed way, and recommended that it be instituted and administered by the OMB, since OMB establishes the strategic allocations of spectrum uses across all Federal applications over time and adjusts the relative consumption of the Federal Government relative to the private sector and the public at large. The Middle Class Tax Relief and Job Creation Act of 2012 requested the OMB to start to do an accounting of Federal systems.[99] Spectrum Currency just extends that accounting to an economic one that could be properly tracked by OMB using a formula that takes into account the new effectiveness spectrum utilization metric described in Section 2.2.

5.4. Redefine the Spectrum Relocation Fund into a Spectrum Efficiency Fund

One of the greatest obstacles to unlocking the productive use of underutilized Federal spectrum bands is the need to cover the very substantial costs of either relocating agency systems and operations or requiring them to share a band. Although relocating a Federal system to a new frequency band is the costliest option, there are also costs associated with expanding shared access to Federal bands, including upfront costs for research, planning, and testing, and for procuring equipment that facilitates band sharing. In addition, irrespective of sharing with non-Federal users, Federal agencies may be able to move to far more spectrum-efficient and/or interference-tolerant technologies, but may have no incentive or authority to do so if their core mission depletes their available budget for more efficient and state-of-the-art radios.

Federal spectrum incumbents need resources to take affirmative steps to enable more intensive access and band-sharing by other users. There is such a potential source: the Spectrum Relocation Fund created by Congress under the Commercial Spectrum Enhancement Act (CSEA) of 2004.[100] Although the Spectrum Relocation Fund facilitated the reallocation of the Advanced Wireless Services band that was auctioned in 2006, the Fund's reimbursements are strictly limited to the actual costs incurred in relocating Federal systems from auctioned bands. The Middle Class Tax Relief and Job Creation Act of 2012 further amended the CSEA to broaden the purposes of the Spectrum Relocation Fund to improve Federal systems left in bands that have been auctioned for commercial use.[101] We certainly applaud Congress for this positive step forward. However, it will be important in the future to improve Federal systems to make spectrum attractive for sharing, but the costs of research, planning, and testing to this end are not reimbursable by the Fund. We recommend that Congress amend the CSEA to broaden the purposes of the Spectrum Relocation Fund further, by renaming it a "Spectrum Efficiency Fund" that can reimburse Federal agencies for general investments in improving spectrum sharing.[102]

Enhancing agency budgets for the purpose of upgrading to state-of-the-art equipment could prove to be a strong incentive, since it would provide agencies with budget dollars above and beyond their normal appropriation. The overall goal would be for the Spectrum Efficiency Fund to be self-financing and budget neutral, allowing market mechanisms (e.g., auctions, user fees) to pay for Federal spectrum evolution over the next decade. The Fund would be replenished from a number of sources related to the Federal Primary Access, Secondary Access, and General Authorized Access allocation regime described in Chapter IV, and ideally would be able to borrow against future auction revenue to help speed the process of transition. The Spectrum Efficiency Fund should be one of the key sources for trading in Spectrum Currency by the agencies (see previous section).

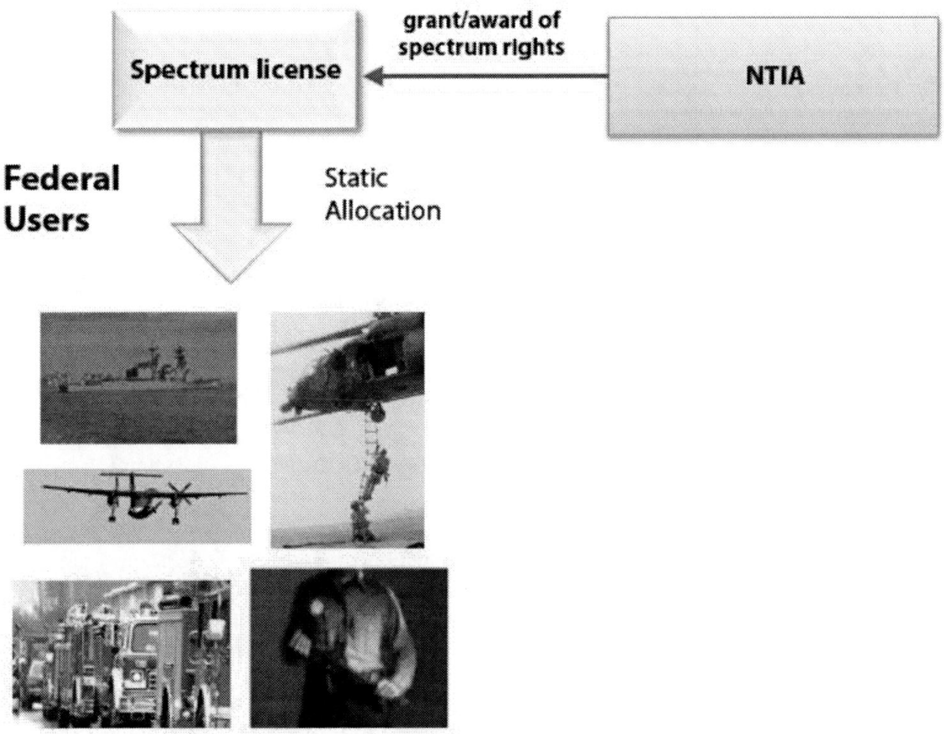

Figure 5.1. (a) Today: Allocations are Static and Determined by NTIA for Federal Spectrum.

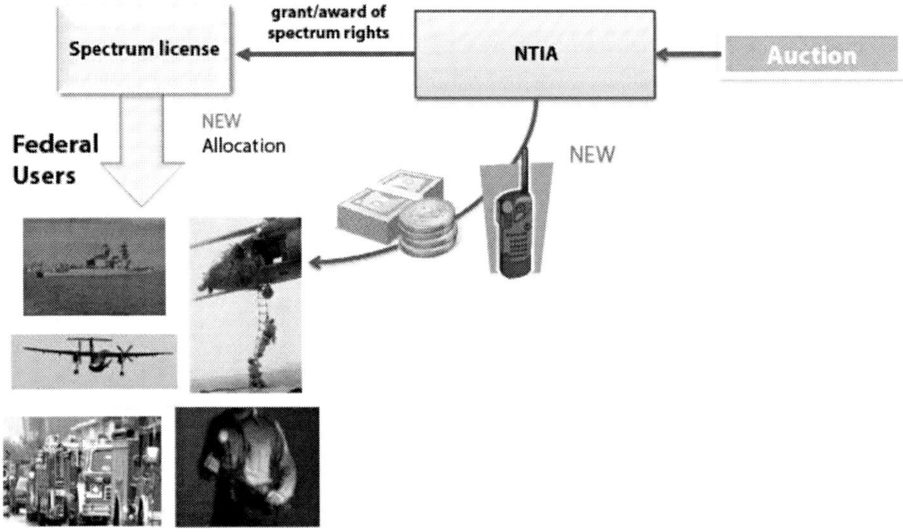

Figure 5.1. (b): Transition to the Future: Auction revenues allow NTIA to provide new spectrum allocations and new equipment for those allocations, enabling bands to be cleared for commercial use.

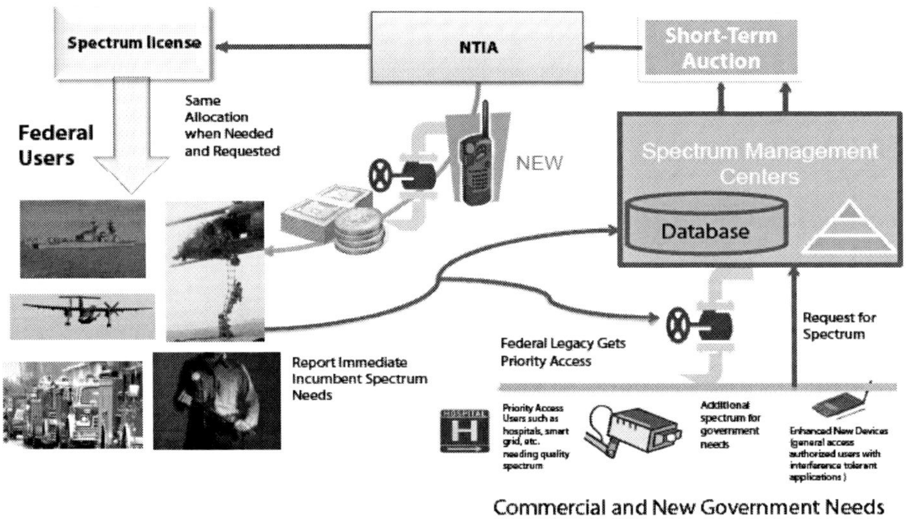

Source: PCAST.

Figure 5.1. (c) Future: Short term auctions can provide temporary access to Federal spectrum that is not being used. Revenue from these auctions allows Federal uses to relocate to new bands or to adopt more spectrally efficient systems in their current bands. The spectrum management center provides allocations to Federal and commercial users depending on spectrum availability and price for spectrum access. Legacy Federal users retain ultimate control over spectrum access.

Figure 5.1. Federal Spectrum System: Today, Transition to the Future, and Future.

One possible OMB method of distribution could be to conduct a competitive auction with Federal agencies using Spectrum Currency to bid for Spectrum Efficiency Fund dollars. The "winners" of this auction end up with current period money, in addition to their normal appropriation, that can be used immediately to make the capital or operating investments needed to implement dynamic sharing for their applications. If they are among the more aggressive in making the transition, they may bet on access to an annual, but limited, amount of funds set aside in the Spectrum Efficiency Fund for these transitions. If they wait, their operating budgets will have to support this transition eventually anyway without incentive budget relief. OMB can play a crucial role in creating opportunities for sharing spectrum by requiring Federal communications systems to be designed so that they can participate in the SAS. The existing OMB Circular A-11 requirement[103] contains some useful guidance on improving the efficiency of procured radio systems, but does not address fundamental design decisions concerning the architecture of systems and possible consolidation of multiple agency systems. Expansion of this guidance to ensure that capability for sharing with both Federal and non-Federal users is an essential design feature would help move the Federal sector away from systems that require exclusive use of spectrum. It would also likely create opportunities to reduce the diversity of communications systems somewhat, and reduce the costs of having many similar but non-interoperable systems being procured. Figure 5.1 explains how this system works today and how it would transition to a new system in the future.

5.5. Spectrum Sharing Partnership Steering Committee (SSP)

Public Private Partnerships (PPPs) are a mechanism that combines public and private sector funding together to reach a common goal. Such partnerships were critical for the development of any number of information technology based industries including the transistor, semiconductors, and GPS. According to one study, approximately two-thirds of award-winning U.S. innovations involve some kind of inter-organizational collaboration such as a PPP.[104]

We believe that spectrum will not be any different. In order for the United States to be competitive in its use of spectrum, a PPP is needed so that government, industry, and academia can work together. Otherwise, it is possible that the U.S. will make this investment and free up spectrum for private use, yet it will not meet industry needs. For this partnership to be successful, however, we believe that a mechanism is needed to prioritize and focus its activities so that its full potential is reached. PCAST recommends that the President establish an advisory mechanism, similar to the Advanced Manufacturing Partnership (AMP) Steering Committee. Based on a PCAST recommendation, President Obama launched the AMP as a national effort that would bring together industry, universities, the Federal Government, and other stakeholders to identify emerging technologies with the potential to create high quality domestic manufacturing jobs and enhance U.S. global competitiveness. Operating within the PCAST framework, the AMP Steering Committee had three targeted outcomes: (1) develop a permanent model for evaluating, prioritizing and recommending Federal investments in advanced manufacturing technologies; (2) recommend a set of partnership projects, focused on advancing high-impact technologies and creating models for collaboration that encompass technology development, innovation infrastructure, and workforce development; and (3) provide recommendations to the administration on the actions required to support investment in advancing manufacturing in the United States.

PCAST recommends that a similar effort be taken on spectrum, specifically that the President appoint an advisory committee of industry executives (e.g. CEOs) to be known as the Spectrum Sharing Partnership Steering Committee (SSP) to advise the SMT on a policy framework and implementation milestones, centered on a public private partnership, for sharing Federally-held spectrum. We expect the SSP to make its recommendations over a one-year timeframe as opposed to being a long-term ongoing effort. As necessary, they may wish to call upon CSMAC and the FCC TAC for technical advice. The products might be short memorandum or reports focused on specific topics the SSP believes are important to address for the effort to be successful.

5.6. International Harmonization

Discussions and research about the need for additional spectrum and spectrum sharing occur not only in the United States, but in other countries as well. As mentioned in Section 1.1, China, Germany, Hong Kong, and South Korea have developed national broadband plans to upgrade their wireless and wire-line broadband platforms at a fast rate. On March 7, 2012, Ed Richards, Chief Executive of Ofcom gave a bold speech in Brussels challenging Europe to move in a new innovative direction that embraces the technologies and policies needed for worldwide leadership themselves.[105] The European Commission just released a report that discussed the topic of shared access to spectrum.[106] A few promising next generation concepts (Box 5.1) are being explored currently in Europe including a proposal from Nokia and Qualcomm called the Authorized Shared Access (ASA)/Licensed Shared Access (LSA) system, the Wireless Access Policy for Electronic Communications Services (WAPECS), and the white space trials in Cambridge, U.K., spearheaded by 17 vendors including Microsoft. Appendix F highlights a number of the spectrum-sharing research projects being conducted in China, the EU, and the United States. A key issue that has arisen is international harmonization. When, as in Europe, many countries are close together, regional if not international harmonization is needed to avoid interference. As indicated in the EU report, only a small amount of spectrum is globally harmonized. Without such harmonization, it will be more challenging for new products to achieve mass market success due to insufficient economies of scale.[107] Other issues leading to the need for international harmonization are interoperability and roaming, spectrum efficiency, and cross-border frequency coordination.[108]

The United States faces a similar challenge in relation to Mexico and Canada, where there is also a possibility of interference if agreements are not reached.[109] As a result, the United States has a number of bilateral and regional agreements in place regarding spectrum.[110] For example, on August 1, 2011, the FCC announced arrangements with Industry Canada and Mexico's Secretariat of Communications and Transportation (SCT) for sharing commercial wireless broadband spectrum in the 700 MHz band along the U.S.-Canadian and U.S.-Mexican border areas.[111] These interim arrangements with Mexico enabled the United States to continue deployments in the 700 MHz band under the FCC's 700 MHz band plan. The FCC also reached an arrangement with Industry Canada for sharing spectrum in the 800 MHz band. In addition, at Mexico's request, the Department of State, in collaboration with the FCC, NTIA, and other Federal agencies, held discussions regarding future use of the 1755-1850 MHz band along the border area in May and August of 2011.[112] Table 5.2 identifies "Fast Track" bands obligated by U.S.-Canada or U.S.- Mexico bilateral agreement(s).[113] A list of many of these agreements can be found in the NTIA Manual of

Regulations and Procedures for Federal Radio Frequency Management. Bilateral discussions with these countries will be needed as the sharing of Federal spectrum in the United States increases. The State Department, advised by the FCC and NTIA, represents the U.S. position at the United Nations International Telecommunication Union's (ITU's) World Radiocommunication Conferences (WRC), which is normally held every three to four years.[114] The purpose of these conferences is to review, and, if necessary, revise the Radio Regulations, the international treaty governing the use of the radio-frequency spectrum. The next conference, WRC-15, will take place in 2015 and will consider spectrum requirements for uses ranging from mobile service allocations for broadband applications to controlling unmanned aircraft from space. One agenda item for that meeting, in response to a U.S. request, is potential new mobile broadband spectrum to support the President's 500 MHz Initiative. Because of the importance of international harmonization for product innovation, interoperability and roaming, spectrum efficiency, and cross-border frequency coordination, the U.S. should encourage international harmonization of spectrum allocations to wireless broadband at the WRC-15, particularly in bands used or planned to be used for wireless broadband applications in the United States, and also work on a bilateral and regional level with its own neighbors, Mexico and Canada.

Box 5.1. Some Leading Spectrum Management Proposals in Europe

ASA/LSA: During the Working Group Frequency Management Meeting in Germany in May 2011[115] of the European Conference of Postal and Telecommunications Administrations (CEPT), representing 48 countries in Europe, a presentation was made on "An evolutionary spectrum authorization scheme for sustainable economic growth and consumer benefit." The work, architected by engineers at Qualcomm and Nokia,[116] enabled the dynamic use of spectrum whenever it was unused by an incumbent user. It deployed cognitive radio techniques, geo-location databases, and sensing when required.

It envisioned secondary shared access that would give cellular 3G and 4G services the ability to initially share with incumbent military (2.3 GHz band) and satellite users (3.8 GHz band) and offer predictable quality of service. Their database system registers the incumbent systems and the secondary system carrier base stations. They were proposing that it be reviewed for implementation throughout Europe. It is being studied by the same standards body reviewing the concepts of white spaces and cognitive radio.

At the most recent meeting of the WGFM in April 2012, they revised and agreed on the most recent report of the Management ASA/LSA system and the members decided to create a WGFM Forum Group or Project Team forward to continue to evolve it and possibly look at it for a possible implementation in the 2.3 GHz band across Europe.[117]

WAPECS: In 2005 the European Commission presented "Wireless Access Policy for Electronic Communications Services" (WAPECS). WAPECS is based on the idea that any communications service might be offered via any platform and that there should be a generic rather than "command and control" regulatory policy and "flexible use" allocations. The approach is purposely being introduced gradually due to international agreements, long-term licensing, and cautiousness about the risk of adjacent band interference. The European Union has a policy that once a band is harmonized, it should stay harmonized, even if greater flexibility can be achieved.

In 2010, CEPT surveyed administrations about their experiences implementing WAPECS. There aren't yet enough proof points to draw conclusions, but it has spawned recent debate in the European Communications Office to consider whether flexibility should be considered as more important, or at least equal, in priority to harmonization so that shared rights can be based on technology that avoids interference, rather than specific allocations.[118]

Cambridge TV White Spaces Consortium: Although White Spaces were first conceptualized and implemented in the United States, the most comprehensive testing has been conducted in Cambridge, U.K., with participation of 17 vendors for more than 10 months, concluding on April 25, 2012.

The Cambridge TV White Spaces Consortium, which comprises leading international companies including, Microsoft and U.K. technology and media companies, successfully demonstrated the potential of television white spaces. The consortium explored and measured a range of applications—rural wireless broadband, urban pop-up coverage and the emerging M2M communication—and found that TV white spaces can be successfully utilized to help satisfy the rapidly accelerating demand for wireless connectivity. The trial analysis found that Cambridge has significant capacity—160 MHz in total, of which 104 MHz were tested for broadband access to rural areas and M2M communications. Geo-location databases provided by Microsoft and Spectrum Bridge were successful as a reliable way to manage frequency use and to quickly adapt to changes in spectrum usage.[119]

VI. The Test City and Mobile Test Service

Finding 6.1: Insufficient opportunities are available to test new architectures, policies, and the new systems proposed in this report for the large scale dynamic sharing of innovative commercial products in the presence of existing real world public safety and Federal incumbent applications.

Recommendation 6.1: PCAST recommends that the Secretary of Commerce, working through the National Telecommunications and Information Administration (NTIA) and the National Institute of Standards and Technology (NIST), provide test services (a Test City and a related Mobile Test Service) to support the development of the policies, underlying technologies, and system capabilities required to support dynamic spectrum sharing. Services would include large-scale sustainable facilities for systems-level testing across multiple frequency bands, including public safety and selected Federal bands. The Secretary should support these services by establishing a Public Private Partnership (PPP) that would pool the resources of Federal, state, and local governments with industry and academia. The Federal contribution to the partnership could be funded, depending on timing and other factors, by NIST's Wireless Innovation Fund, by the Public Safety Trust Fund, and potentially by the Office of the Secretary of Defense and the National Science Foundation.

The new spectrum architecture advocated in this report has enormous potential to free up spectrum capacity and unleash a wave of technological innovation. For the architecture to be embraced, however, incumbent spectrum users need to have confidence that dynamic sharing of the spectrum will not cause harmful interference to their existing systems. New entrants need to be sure of the reliability of spectrum access that their business models depend on and need a well-defined, streamlined, process for testing and approval of new devices.

This report therefore calls for the creation of an urban Test City complemented by a Mobile Test Service to support rapid experimentation and gain essential operational test data to establish the dependability of both the technology and the management techniques supporting the new architecture. In the first stages of implementing the new spectrum architecture, these test facilities will be essential for implementing the 1,000 MHz Super Highways proposed in this report, and also for assessing options for sharing public safety spectrum services with commercial users.

6.1. Steps Needed to Institute Spectrum Sharing

The process of bringing new dynamic spectrum sharing devices to market starts with the identification of spectrum for sharing and ends with certification of dynamic sharing devices. Recently both 700 MHz whitespace devices and 5 GHz Unlicensed National Information Infrastructure (UNII) devices have been approved for dynamic spectrum sharing. Both of these efforts took many years to reach commercial viability, which is not unexpected because in many ways they were breaking new ground in spectrum sharing. There are many lessons to be learned in planning and implementing the testing approaches that will be used. One of the better recent examples of successful unit testing for the UNII band was by the FCC Labs.[120]

However, this process must be shortened to attract significant commercial investment in dynamic sharing systems. Once spectrum for sharing has been identified, we propose the following process for bringing new dynamic spectrum sharing devices to market:

Identification and characterization of incumbent systems in target sharing band: Characterization of the incumbent systems in some region of spectrum is a necessary first step in establishing a shared band, and is done by government laboratories since some of the incumbent Federal systems may be classified. Today's static database for recording Federal spectrum assignments means that identification of incumbent systems can be time consuming even in the simplest cases. The SAS described in this report will largely automate and greatly speed this process. The characterization of the incumbent systems in the 5 GHz band performed by NTIA's Institute for Telecommunication Sciences[121] provides an excellent example of the characterization work that needs to be done in each of these bands.

Education of any interested commercial parties on the characteristics of incumbent systems: Commercial vendors proposing to design dynamic sharing systems must understand the Federal systems with which they hope to share spectrum. Because Federal and commercial systems are often quite different, vendors must first gain a good understanding of what types of interference might cause problems for the incumbent systems. A good first step would be for the laboratory doing the characterization to hold an industry conference to disseminate its results and start a dialogue on potential sharing approaches.

Identification of viable sharing approaches: The dialogue between the commercial parties and the labs tasked with characterizing the incumbent systems to develop possible sharing strategies is not a standardization step but a more flexible process in which promising approaches can be better defined and evaluated.

Testing of initial prototype dynamic sharing devices: This step identifies incorrect assumptions in a proposed sharing method and pinpoints elements that need further refinement. The conclusion of this stage is finalization of the method, which may be accompanied by a parallel standardization process.

FCC lab testing of commercial devices for approval: Before commercial devices can be put into operation, they will be certified by the FCC Labs to insure that they meet the identified requirements to prevent harmful interference to incumbent government systems.

Ongoing evaluation, spot checks of devices on market and troubleshooting: It will be important to have a mechanism to identify and resolve performance issues that arise once spectrum is actively shared, as operational problems will likely arise that were not discovered in the preceding steps. In the UNII band, for example, such cases occurred after devices were in the market, but field testing and evaluation were able to identify the issue and define a remedy.

The existing government test labs and facilities, as described in Appendix G, have the capabilities to support most of this process today. We have identified two areas where additional capabilities can speed up and improve the process: a Mobile Test Service and a scale testing facility which we call a Test City.

6.2. Mobile Test Service

The different requirements of the incumbent Federal users and the potential commercial users lead us to propose a mobile testing capability complementary with existing testbeds. The existing testbeds provide capabilities to test against incumbent systems that are local or can be brought in on a temporary basis. Unfortunately many of the incumbent systems such as radars or ship platforms cannot be easily moved to the testbed. Existing facilities have limited support some of this today. For example, the Naval Research Laboratory (NRL) provides testing against certain ship platforms and the NTIA's Institute for Telecommunications Sciences (ITS), in Boulder, Colo., has packed up equipment and moved it to locations such as San Diego and Puerto Rico for testing against incumbent systems.[122] However, a comprehensive Mobile Test Service, with support for numerous bands and power levels, will be necessary to cost-effectively show the utility of sharing with the numerous legacy systems from multiple government agencies.

To test the potential interference and the operational characteristics of dynamic sharing systems in the presence of incumbent Federal systems, we propose a standalone Mobile Test Service that can be moved to different locations where the relevant incumbent systems are deployed. These may be remote rural locations where specific legacy military systems are in operation, or coastal areas where access to specific naval platforms can be obtained. Mobile testbeds can also be used to investigate rural applications such as communications for rural transportation systems or smart agriculture or environmental monitoring. They may also be sent to complement existing testbeds in different regions.

The Mobile Test Service we envisage will be similar to a fleet of "cells on wheels," or COWS, as they are known in the cellular industry (see Figures 6.1). The Mobile Test Service vehicles would be outfitted with equipment shelters, power generation, telescoping antenna masts, backhaul network interface equipment, and instrumentation to monitor and measure the dynamic spectrum sharing experiments. We call this a Mobile Test Service rather than a testbed because it embodies the infrastructure for the mobile test system but not the spectrum sharing equipment itself. A typical use case would be for a testbed, after performing initial tests in their environment, to travel to the location where there is heavy incumbent use of the frequency bands of interest. Given that we see an increasing need for testing against incumbent systems in various locations, the goal of the mobile test infrastructure is to enable the testbeds to perform remote experiments more easily and more cost effectively. Such a testbed could include a number of handheld software radio nodes that can be used to sense the spectrum to avoid the "hidden node" problem of sensing based systems and to create test waveforms for different spectrum access approaches.

Source: www.pscr.gov/about_pscr/highlights/700mhz_demo_net_032012/ps_700mhz_ spring_ 2012_ stakeholder_mtg_info.php.

Figure 6.1. Mobile "Cell on Wheels," an element of the Department of Commerce Public Safety Communications Research (PSCR) Public Safety Broadband Demonstration Network.

Beyond the significant capabilities provided by the Test City facility, the companion Mobile Test Service will enable specific tests to be undertaken in environments that are not available in the urban domain of the Test City. This ability to take the test capability that is roughly comparable to the capabilities of the Test City "on the road" is a strong differentiator from existing facilities. Together the Test City and Mobile Test Service will offer capabilities that do not currently exist anywhere in the world. This should be of enormous value in enabling U.S. leadership in the coming age of dynamic spectrum sharing systems.

6.3. Creating the Test City

Historically, experimental deployments, commonly called "test beds" (see Figure 6.2) or "pilots," allow technology to be tested under simulated real-world conditions, but almost exclusively in an isolated and controlled environment.

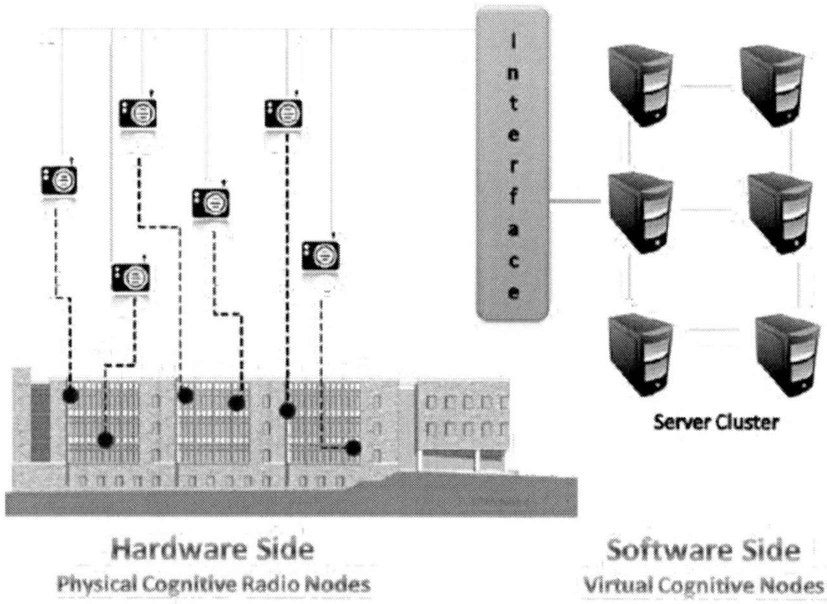

Source: Newman, T.R. "Cognitive Radio Network Testbed (CORNET)." PowerPoint, Wireless @ Virginia Tech. Slide 9.

Figure 6.2. Indoor Cognitive Radio Network Testbed (CORNET) used at Virginia Tech for Radio Resource Management Research. The test bed consists of 48 agile radio nodes to test dynamic spectrum access, spectrum security, location dependent radio management and policy (radio environment maps), and other technologies.

Real-world testing of dynamic sharing principles and the evolving technologies supporting them are necessary to provide the basis for wider deployment and to develop shared spectrum methods, standards, technologies, and trust[123] mechanisms that would make dynamic sharing of unused and underutilized spectrum capacity scalable. As systems grow in complexity, interconnectedness, and geographic distribution, they increasingly experience emergent behavior. Emergent behavior is that which cannot be predicted through analysis at any level simpler than that of the system as a whole.[124] Large software and network systems demonstrate emergent behavior and the need to test for these types of behavior is now well understood. The combination of dynamic spectrum sharing and small cell architectures will scale wireless systems into the domain where we expect emergent behaviors to occur. Emergent behavior is not necessarily bad—for example, a colony of ants demonstrate greater capabilities and intelligence than individual ants—but understanding these emergent behaviors is critical to enable successful large scale system deployments for both Federal and non-Federal users.

In connection with public safety systems, for example, the Visiting Committee on Advanced Technology of the National Institute of Standards and Technology (NIST) has said that "test beds will be vital for the exploration of new technology, methods, ideas and architectural enhancements. It would be a major mistake to imagine that the design of a public safety system is a one-time event. It will be part of a continuing evolution of telecommunication and information technology and will play a key role in facilitating that evolution."[125] Similarly, Federal systems need real-life experience, in real environments, in co-existing with other users of the spectrum before the technologies can be broadly deployed.

The introduction of new wireless technologies and capabilities in both the commercial and government spaces faces several challenges: users are not generally aware of the possible benefits; are reluctant to embrace new and unproven technology; face budget constraints in upgrading or replacing equipment; and have concerns about the maturity and availability of new systems. Experience from the Test City can alleviate all of these concerns.

The primary goal of the proposed Test City is to facilitate and accelerate the transition to a new architecture, leveraging available technology to open up opportunities for dynamic spectrum sharing as soon as possible. What is needed is a robust infrastructure located in an urban environment where the choice of frequency, standard, protocols and sharing strategy can be modified independently and dynamically, and tested together with the spectrum management tools and enforcement mechanisms. To be clear, Test City activities would take place after the initial testing and verification in the labs has been performed.

Unlike traditional wireless network testbeds, the Test City will be an urban facility with access to a broad set of spectrum bands where the systems under test will experience the rich, ever-changing characteristics of a challenging real world environment. The Test City facility will have the low level functional capabilities to assign technology mechanisms, receiver management protocols, and enforcement technology in addition to actual radio and network operation.

The Test City must support the architecture's transition to broader bands by adding support for new bands as they become available for sharing and must be able to test systems that operate over a very large frequency range. It must also support large scale experiments using among other approaches small cell architectures to validate both the utility of higher frequency spectrum and the increased spectrum efficiency of this architecture. At the higher level, the Test City facility must support new system structures, management approaches and importantly new applications that exploit the opportunities provided by the expanded availability of spectrum inherent in the dynamic sharing approach.

Finally, the Test City must support experimentation with new economic models with the goal of developing financially desirable dynamic sharing mechanisms to assure proper incentives to transition to the new approach.

In developing the design for the Test City, there are two key constituencies with paramount concerns: incumbent public safety and Federal users, and entrant Federal and commercial sector users who wish to share the incumbent spectrum (see Box 6.1). Safeguarding incumbent users is the first priority, as they will not be comfortable moving forward without quantitative data from field trials that demonstrate that the new dynamic sharing systems will not create harmful interference to incumbent systems. Conversely, entrant Federal and commercial users need to understand that the restrictions placed on the dynamic sharing systems to safeguard incumbents will not preclude the provision of viable new commercial services.

Without these assurances entrants will be unwilling to commit significant investments in time, energy, and dollars required to implement these systems and will instead view dynamic shared spectrum systems as an unviable commercial alternative to traditional exclusively licensed spectrum based systems.

The Test City will be used to test the operational SAS created by the various third-party providers, determine appropriate receiver interference limits, and validate the technical enforcement mechanisms. It will enable migration to the long term architecture based on large open spectrum regions. It will also be used to enable the testing of new systems and applications in a real world environment to enhance the rate at which these systems and applications can be properly developed and deployed to meet the needs of the U.S. citizenry and to enable the economic enrichment of the developers and the nation.

Appendix G provides additional details on the Test City and Mobile Test Service, and their estimated cost. We believe that the test city could be funded through a competitive process based on pre-determined technical considerations. Regional clusters of local industry associations, government, and academia, whose cities meet those characteristics, could then develop a proposal to host the test city in their region. This model has been successful in the past as a way for a region to leverage their own innovation investments as the presence of suppliers, information, and role models that creates a favorable environment for innovative spin-offs.[126]

6.4. Summary

The Test City will provide a critical multi-spectrum broadband urban platform enabling the United States to be the first mover in the competitive drive to design and deploy a state-of-the-art, next-generation dynamic spectrum sharing based communication network. It and the mobile facility will enable the validation of new technologies for future commercial use, future-proof the way spectrum is managed and regulated, and allow the exploration of emergent behaviors that will occur as large scale, small cell, dynamic spectrum sharing systems are deployed. If the United States moves quickly in the development and testing of disruptive dynamic spectrum access technologies and systems, it can dominate the next generation of wireless devices. If not, others will clearly capture this initiative and even our current wireless position will be lost.

The dynamic spectrum testing facilities would facilitate a new wireless eco-system that would include not only the engineering talent to develop the hardware and software systems, but also people who can manufacture, install and maintain such equipment. Education and training should therefore be an integral part of the dynamic spectrum Test City development. The regulatory community should also be involved, to help minimize business risks by rapidly providing the proper regulatory framework and supporting the experimental fielding of the technology.

Perhaps the most important aspect of the facilities will be the opportunity for wireless system and applications developers to test the capabilities of their products and systems in complex "real world" environments. More importantly, the tremendous number of companies that rely on reliable wireless technologies would gain from more capable technology base.

Box 6.1. Dynamic Spectrum Sharing for Public Safety

The Middle Class Tax Relief and Job Creation Act of 2012 reallocates the 700 MHz D block spectrum to public safety and provides $7 billion in funding to build a nationwide public-safety broadband network. However, $7 billion is not enough to build a complete nationwide network. The bill allows for leasing of the spectrum to secondary users, with any revenue gained from such a leasing agreement being used for constructing, maintaining, operating or improving the radio access network. The SAS described in this report provides a means to implement this secondary access, generate revenue, and enable public safety to retain usage of the spectrum in times of need as the primary user. Public safety would be the primary user of the spectrum. Secondary access, as administered through the SAS, would be granted via certificates with a finite time-to-live. The secondary user devices would have to constantly renew their certificates, allowing public safety to quickly reclaim exclusive use of the spectrum in times of emergency. This would be the most cost-effective way for public safety to generate revenue from secondary use of their spectrum, since it would leverage the spectrum access infrastructure built for sharing of other federal bands. This approach could also generate revenue for public safety prior to construction of the FirstNet network, as the secondary access could be provided before the network is built and provide additional funds for the construction of the network. HR3650 also calls for public safety to give back the T-Band spectrum (470 MHz – 512 MHz). This spectrum is primarily used by public safety agencies in large cities across the nation, and many cities have recently invested a significant amount of money in T-band infrastructure. Dynamic spectrum sharing provides an alternative to public safety reallocating out of the T-band. Instead, this spectrum could be made available for sharing with the incumbent public safety systems as the primary users. The T-band also includes business/industry and Specialized Mobile Radio (SMR) operators in 13 of the largest U.S. cities. Spectrum sharing across this band provides a way to harmonize usage of the T-Band rather than have public safety relinquish the spectrum while commercial users stay in place. Finally, dynamic spectrum sharing provides benefits to public safety far beyond what the 700 MHz public safety broadband allocation can provide. This report identifies 1 GHz of spectrum to be initially made available for dynamic spectrum sharing. Through the SAS, public safety can gain access to additional spectrum that they are unable to access today. The device ecosystem driven by this spectrum being made available for commercial sharing will result in affordable devices for public safety in this band, and the gigahertz of spectrum would provide tremendous additional communications capacity for public safety in both emergency and non-emergency situations.

VII. Summary and Implementation Plan

Finding 1.1: Spectrum provides a great economic opportunity for the Nation. The economy created by making spectrum abundant has the opportunity to provide social benefits of over $1 trillion and millions of jobs for Americans over many years. Most importantly, it will provide a foundation for American economic and technological leadership.

Finding 1.2: Clearing and reallocation of Federal spectrum for exclusive use is not a sustainable basis for spectrum policy due to the high cost, lengthy time to implement, and disruption to the Federal mission. Sharing of Federal spectrum, however, would provide the basis for economic and social benefits for the Nation.

Finding 1.3: The fragmented partitioning of Federal spectrum leads to inefficiency, artificial scarcity, and constraints on current and future Federal and non-Federal uses.

Recommendation 1.1: PCAST recommends that the President issue a new memorandum that states it is the policy of the U.S. government to share underutilized Federal spectrum to the maximum extent possible that is consistent with the Federal mission, and requires the Secretary of Commerce to immediately identify 1,000 MHz of Federal spectrum in which to implement the new architecture and thereby create the first shared-use spectrum superhighways.

Flexible and affordable access to spectrum is essential to continuing economic growth in the United States, to the maintenance of secure and reliable wireless services for national security and public safety uses, and for many other essential Federal and local services. This report lays out an ambitious but achievable program that can transform spectrum scarcity into abundance by making sharing rather than exclusivity the norm for spectrum access. Taking this step represents a continuation and an expansion of the President's Directive of 2010, and will set the United States on a path to maintain economic growth and technological leadership. Moreover, carrying out this report's recommendations on Federal spectrum will enable the United States to establish a leading position in technologies that will ultimately be deployed more widely across the spectrum and in other parts of the world, as the need for spectrum capacity forces evolutionary change in spectrum management practices globally.

This chapter describes steps that should be taken to implement the recommendations of this report in a coherent way. The following sections describe an implementation strategy for each of the report's main recommendations.

7.1. Adopt a New Spectrum Management Architecture

Finding 2.1: Sharing of Federal spectrum provides an opportunity to deploy a wholly new approach to Federal spectrum architecture and policy by establishing large shared spectrum blocks, new effectiveness metrics, and coordinated and prioritized Federal and commercial use.

Finding 2.2: Wireless architectures have evolved from a single model of high-power, high altitude base stations to a mix of capabilities, ranging all of the way from base stations to offload onto commercial Wi-Fi. This provides an opportunity to locally exploit Federal spectrum sharing opportunities that would not be otherwise compatible with high power operations (such as LTE).

Recommendation 2.1: The Secretary of Commerce, in collaboration with the Federal Communications Commission (FCC), should establish a mechanism to provide the Federal Government with the ability to manage the sharing of Federal spectrum.

Federal spectrum should be divided into substantial frequency blocks with common characteristics, rather than the current narrow band service-specific static allocation scheme. In addition, rather than the current pre-allocation and assignment of spectrum, there should be a new "dynamic sharing" model that makes spectrum sharing by Federal users the norm, and also allows sharing with commercial users. Shared access to Federal spectrum should be governed according to a three-tier hierarchy: Federal primary systems would receive the highest priority and protection from harmful interference; secondary licensees must register deployments and use in a database and may receive some quality of service protections, possibly in exchange for fees; and General Authorized Access users would be allowed opportunistic access to unoccupied spectrum to the extent that no Federal Primary or Secondary Access users are actually using a given frequency band in a specific geographical area or time period. All Federal agencies should be required to cooperate in the implementation of these changes.

Recommendation 2.2: The Secretary of Commerce, working through the National Telecommunications and Information Administration (NTIA) and the FCC, should authorize and implement, directly or through commercial providers, a Federal Spectrum Access System (SAS) to serve as an information and control clearinghouse for the band-by-band registrations and conditions of use that will apply to all users with access to each shared Federal band under its jurisdiction. The SAS will protect Federal operations from interference while allowing non-Federal users to access underutilized spectrum in Federal bands. Underutilized spectrum capacity in Federal bands should be made available to the greatest possible extent for non-interfering shared use, based on the principle that exclusive assignments should not be taken as a justification for letting unused or underutilized spectrum lie fallow.

Initial Steps to Implement a Spectrum Access System

Sharing access to the underutilized capacity in all Federal spectrum bands will be a major undertaking, although one where the enormous economic benefits will dwarf the very modest additional costs to NTIA and Federal spectrum users. NTIA estimates that it currently has records on 250,000 frequency assignments to Federal departments and agencies in its 30-year-old Master File. The Master File is a static administrative database that is in many cases out of date or lacking the technical information that the proposed Spectrum Access System would need to determine transmit permissions and the terms of use for shared access based on frequency, location, time and other variables.

NTIA is currently in the process of collaborating with the Defense Department on the design of a new, automated Federal Spectrum Management System (FSMS). We recommend that this effort should simultaneously collect and update whatever additional data on Federal assignments and operations is needed to implement a parallel (and unclassified) Spectrum Access System database to govern spectrum sharing. Because of the large number of frequency assignments and the need to collect additional data, implementation must realistically begin by prioritizing bands on a cost-benefit basis. The bands already identified by NTIA for Fast Track evaluation should be among the first that can be profiled in the SAS database and coordinated with the FCC for promulgation of rules specifying the particular band-byband "terms of use" for private sector sharing. In particular, the NTIA identified a contiguous 950 MHz, between 2700 MHz and 3650 MHz, which has substantial underutilized capacity and potentially could be opened for sharing on a secondary and/or general authorized access basis. Within this priority range, implementation can be further prioritized based on bands that are both underutilized and occupied by Federal primaries that present fewer technical hurdles to dynamic sharing. As mentioned earlier, for example, the NTIA has identified the 3550–3650 MHz radar band as one that could be shared outside of specified exclusion zones, the size of which would vary dramatically depending on the power levels and antenna heights of the secondary and tertiary users. Indeed, the Chairman of the FCC has similarly suggested that this band be reviewed for small cell use.[128]

Because the FSMS is intended to house both classified and non-classified data, NTIA and DOD should develop the new system from the beginning to create cross-domain access controls that permit a certified database manager to interface with the FSMS and to access non-classified data directly, but only summary or filtered access to classified information. NTIA could also leverage the SAS and its spectrum assignment registration requirement as a means of automating the process of agency assignment, and make it much easier to support civil use of Federal spectrum on a secondary basis, as well as enabling agencies to determine the reality of actual band use and availability.

As one band after another is prioritized by NTIA for representation in the SAS database, the FCC should simultaneously open a notice of proposed rulemaking to determine the "terms of use" for the band, including what secondary access is appropriate and the rules governing general authorized access. We expect that this process could be similar to the final stages of the White Spaces rules and might require a period of device testing. If the NTIA and FCC determine that shared access to a band can be managed through the existing TV Bands Database, or some other authorized third-party database, we expect that any associated costs would be paid by the private parties using the spectrum. Like the White Spaces for the TV Bands Database, the FCC can authorize the database administrators to recoup any costs through fees charged to device makers and/or device users. We expect that at least for general authorized access this would be a relatively small one-time fee collected in volume from the handful of equipment manufacturers or others that sell devices certified to operate in shared bands.

7.2. Technology Advancements and Challenges to Solve

Finding 3.1: Spectrum management and regulation is focused on the characteristics of transmission, whereas receiver characteristics increasingly constrain effective and flexible spectrum usage.

Recommendation 3.1: The Secretary of Commerce working through the National Telecommunications and Information Administration (NTIA), in cooperation with the Federal Communications Commission (FCC), should establish methodologies for spectrum management that consider both transmitter and receiver characteristics to enable flexible sharing of spectrum. To safeguard primary Federal users, FCC should require that future non-Federal devices will be permitted to share government spectrum as Secondary Access users only if they are certified to operate within the stated interference limits for the band of interest. Initial specification of protection should be reviewed such that they safeguard new FCC assignments against harmful interference while grandfathering in existing devices and operations.

Initial Steps to Implementing a Receiver Regulation Framework

Due to the differences in legacy usage, as well as the different objectives for Federal Government and non-Federal commercial spectrum use, receiver management must be addressed differently by the NTIA and FCC. There is already considerable sharing between Federal users today, both in Federal-only bands and in bands shared with non-Federal users.[129] In the 1755-1850 MHz band, for example, 19 agencies operate 10 types of systems, with 3,183 individual assignments.[130] The sharing is enabled by defining unique operating regions along the traditional dimensions of space, time and frequency. In order to expand usage by enabling more dynamic sharing methods, it will be necessary to carefully define and manage the receiver operations in these bands and at the adjacent edges of the allocations. In order to facilitate more intensive and efficient sharing among Federal users, the NTIA should set and publish receiver interference limits using a transparent process for government assignments. This will facilitate the deployment of automated assignment tools.[131]

Within non-Federal licensed bands where the transmitters and receivers are under the control of licensees that use similar technology, have symmetrical interests, and interact with each other repeatedly, economic incentives ensure proper receiver operation. For example cellular systems are all designed to work with the expected interference from other cellular systems within the band. This allows the carriers to improve their spectrum efficiency within the band by maximizing the number of users, given the interference environment. The primary problem in non-Federal spectrum is at the edges between spectrum allocations. In new allocations, the FCC would follow the same procedure as recommended for new NTIA assignment. The FCC, NTIA, and the White House SMT will need to work together to specify the interference environment to be tolerated by non-Federal devices sharing with existing incumbent Federal users within each band, and that the FCC specify the appropriate device requirements that will ensure adequate operation of non-Federal devices.

In cases where the NTIA or FCC has decided to go beyond setting interference limits, e.g. by prescribing minimum front-end filter requirements for receivers, it may tighten these requirements from time to time. However, obtaining the maximum value from radio operations is a system optimization challenge, both in engineering and economics. It will not always be the case that the optimum solution requires improving receiver quality; it may be most efficient to deploy poorer receivers if an operator obtains bands to choose from through spectrum sharing, and can cope with interference degradation by hopping to another channel.

In the immediate term, we see that the NTIA will develop a framework for Federal receiver management, including:

- Specify receiver interference limit parameters (how to define signal strength profiles, required granularity in spatial, temporal and frequency parameters, use of measurement vs. modeling to resolve disputes, etc.), and device performance mandate options (applicability and use of self-certification vs. mandated device performance standards)
- For new Federal assignments, define protocol for defining receiver interference limits
- Determine high-priority bands where receiver interference limits should be deployed first. Requirements and roll-out sequence for intra-Federal sharing, Federal/non-Federal sharing.

- Decide what additional trust building mechanisms are required for unlicensed consumer devices sharing with Federal enforcement, e.g. manufacturer warranties for Part 15 (unlicensed) certification vs. FCC-defined receiver standards, including access to the dispute resolution mechanism previously discussed.

Also in the immediate timeframe, the FCC should begin the Notice and Comment cycle on implementing receiver interference limits as part of license terms for new allocations, updating old licenses to include receiver interference limits, and ex ante enforcement mechanism for non-Federal devices sharing with Federal users. Congress should begin the legislative cycle for changes in the Communications Act required by new receiver regulation management approach.

7.3. New Application Economy

Finding 4.1: Moving to a dynamic sharing model for Federal spectrum would unlock economic benefits by allowing the private sector to make intensive use of currently underutilized parts of the radio spectrum. A well-designed Federal spectrum policy opens up opportunities for innovation and growth in sectors that are barely imagined, much less well-defined, when the policy choice is made.

Finding 4.2: Sharing of Federal spectrum provides an opportunity to deploy new spectrum management principles such as shorter term licenses that would be appropriate to new and innovative spectrum-based services and products. This provides an opportunity to collect revenue to the Treasury from the private sector for assured use of spectrum.

Recommendation 4.1: PCAST recommends that policies enabling commercial access to Federal spectrum be based primarily on their effects on innovation and growth in wireless devices, services, and associated markets; direct revenue considerations should be treated as secondary. The Office of Management and Budget (OMB) should develop a model to assess future economic growth effects of wireless allocations as well as revenue from increased economic activity.

The National Telecommunications and Information Administration (NTIA) and the Federal Communications Commission (FCC) should continue to embrace the current proven unlicensed model and, for licensed spectrum, explore adding new short and medium-term spectrum license models that could both foster growth for these new applications and collect revenue.

Implementation

A variety of license model approaches to dynamic sharing of spectrum between users, in our case Federal and non-Federal users, is not only practical, but may be necessary if we want to motivate the development of all these new innovative applications over the next decade.

Many of those applications may desire or even require quality of service level predictability to create effective markets for their products. To have that environment available, we believe that the NTIA and FCC should develop and experiment with a range of approaches to sharing, including, in particular, the already proven approach based on unlicensed access, combined with new intermediary models of short or medium term prioritized use and those models can choose to collect usage fees. These models can be easily added to the SAS once it is implemented by the third party providers and those same providers would then also act as billing partners to OMB for the spectrum usage fees, generating both revenue for the SAS private sector providers to invest in their systems and for the Federal Government.

7.4. Starting with Federal Spectrum

Finding 5.1: There is no incentive system today for Federal Government agencies to be efficient in their use of spectrum or to share spectrum allocated to them with the non-Federal sector.

Finding 5.2: A public private partnership (PPP) is the best mechanism to ensure that optimal use is made of the Federally-held spectrum and of related investments in spectrum research and testing.

Finding 5.3: International harmonization of spectrum policies is essential to product innovation, interoperability and roaming, spectrum efficiency, and cross-border frequency coordination.

Recommendation 5.1: PCAST recommends that the White House Chief Technology Officer (CTO) with senior officials at an equivalent level from the National Security Staff (NSS), the Office of Management and Budget (OMB), and the National Economic Council (NEC) formalize a Spectrum Management Team (SMT) to work with the National Telecommunications and Information Administration (NTIA), the Federal Communications Commission (FCC), and the major Federal agencies that use spectrum to carry out the President's directive.

Recommendation 5.2: PCAST recommends that the NTIA, working with the SMT and Federal agencies, reexamine the partitioning of Federal spectrum usage in light of current and emerging technology. One objective of this reexamination is to aggregate current spectrum partitions to create substantial frequency blocks in order to facilitate sharing through common technical use rules.

Recommendation 5.3: PCAST recommends that the President indicate that all Federal agencies should cooperate with the SMT and NTIA to establish and implement a government-wide process and mechanism to share Federally-held spectrum. Within one year, the SMT working with the NTIA should formulate concrete 5-year and 10-year goals for Federal spectrum sharing opportunities in order to recommend to the President how to appropriately update his 2010 goal of making 500 MHz of Federal and non-Federal spectrum available over the next 10 years.

Recommendation 5.4: PCAST recommends that OMB, working with the SMT and NTIA, take steps to implement a mechanism that will give Federal agencies incentives to share spectrum. Such a mechanism would accurately internalize the opportunity cost of Federal spectrum resources and manage them over long time horizons using a "currency-like" accounting, allocation, and incentive system ("Spectrum Currency").

Recommendation 5.5: PCAST recommends that OMB should implement a sustainable funding mechanism to foster a Federal spectrum sharing system. The existing Spectrum Relocation Fund should be redefined as a revolving "Spectrum Efficiency Fund" that recycles private sector payments for use of Federal spectrum into reimbursements to Federal agencies for investments that facilitate spectrum sharing and enhance spectrum efficiency. Congress should allow the Fund to reimburse qualifying costs by any Federal service, not just those in revenue-generating bands.

Recommendation 5.6: PCAST recommends that the President appoint an advisory committee of industry executives (e.g. CEOs), to be known as the Spectrum Sharing Partnership Steering Committee (SSP), to advise the SMT on a policy framework to maximize commercial success, centered on a public private partnership for sharing Federally-held spectrum, and implementation milestones that lay the groundwork for the first spectrum superhighways.

Recommendation 5.7: The United States, represented by the Department of State with advice from NTIA and the FCC, should make international harmonization of spectrum allocations to wireless broadband, particularly in bands used or planned to be used for mobile broadband applications in the United States a key element of the U.S. position at the 2015 World Radiocommunication Conference (WRC-15) and in bilateral and regional discussions with its own neighbors, Mexico and Canada.

Implementation

Federal spectrum provides a unique opportunity for the United States to meet exponentially increasing spectrum demand and establish a leading position in the technology that will ultimately be deployed worldwide as all countries of the world are forced to adopt more aggressive spectrum sharing as a necessary policy. The cornerstone of this new approach is the creation of very large regions of spectrum.

There is the potential to open up as much as 1,000 MHz of shared Federal/non-Federal spectrum for small cell, low-power secondary access and general authorized access. Making such a wide expanse of spectrum available can stimulate needed private sector investment in mass-market technologies, devices and services designed to operate on either a secondary or general authorized access basis using dynamic spectrum access to harvest underutilized capacity in these Federal bands.

Carrying out the President's Agenda will require the re-involvement of the White House, an incentive system to motivate the Federal agencies to share, and a funding mechanism to retrofit and improve the effectiveness of their equipment and reward agencies who are early adopters.

Spectrum currency is our name for a synthetic currency that would give agencies a means to identify the opportunity costs associated with their use of spectrum and to obtain benefits by sharing or vacating some parts of their assigned spectrum and provide a way for them to "buy" their spectrum usage rights and reduce their spending by improving spectrum efficiency. The 2012 payroll tax agreement requested the OMB to start to do an accounting of Federal systems. Spectrum Currency just extends that accounting to an economic one that should be administered by OMB.

To turn their gains in efficiency to practical advantage, agencies desiring to accelerate their transition to the new scheme could use their spectrum currency to bid every year for equipment credit from the Spectrum Efficiency Fund (an evolution of the current Spectrum Relocation Fund) that would enable them to increase their service quality.

7.5. The Test City and Mobile Test Service

Finding 6.1: Insufficient opportunities are available to test new architectures, policies, and the new systems proposed in this report for the large scale dynamic sharing of innovative commercial products in the presence of existing real world public safety and Federal incumbent applications.

Recommendation 6.1: PCAST recommends that the Secretary of Commerce, working through the National Telecommunications and Information Administration (NTIA) and the National Institute of Standards and Technology (NIST), provide test services (a Test City and a related Mobile Test Service) to support the development of the policies, underlying technologies, and system capabilities required to support dynamic spectrum sharing. Services would include large-scale sustainable facilities for systems-level testing across multiple frequency bands, including public safety and selected Federal bands. The Secretary should support these services by establishing a Public Private Partnership (PPP) that would pool the resources of Federal, state, and local governments with industry and academia. The Federal contribution to the partnership could be funded, depending on timing and other factors, by NIST's Wireless Innovation Fund, by the Public Safety Trust Fund, and potentially by the Office of the Secretary of Defense and the National Science Foundation.

Implementation of the Test City

The Test City should be contracted by NIST, with help of the NTIA and FCC, with the purpose of testing this report's recommendations as they apply to public safety and various Federal bands, initially focusing on the 1,000 MHz chosen region, which we believe will be the 2700 to 3700 MHz region. If so, we believe those tests could start in the 3550–3650 MHz band and then be extended to the broader bands comprising the 2700–3700 MHz range to begin to test the concept of systems that operate over a very large frequency range. Over an even longer time horizon, the spectral environment should be extended both up and down the spectral region to a variety of interesting bands as they are identified as desirable areas of exploration for dynamic sharing opportunities.

Implementation of the Mobile Test Service

The Mobile Test Service is focused on reducing the time and cost associated with testing against incumbent systems, such as radars or ship-based systems, that cannot be brought to an existing test facility. The Mobile Test Service would be a resource for existing government test labs to utilize to facilitate and accelerate the validation of new sharing systems. It would be a fleet of vehicles that test facilities can load equipment and systems for testing into. The vehicles contain a shelter for equipment, telescoping masts, power sources and measurement equipment. Since testing against incumbent systems is an early requirement in the process of bringing new devices to market, funding the Mobile Test Service should be an early priority.

7.6. Immediate Selective "General Authorized Access" Sharing

> **Finding 7.1:** Expansion of the white space system to include certain space-to-ground and radar-based Federal bands could allow immediate "general authorized access" device usage while the other recommendations of this report are being enacted.
>
> **Recommendation 7.1:** PCAST recommends that the Federal Communications Commission (FCC), working with the National Telecommunications and Information Administration (NTIA) and the Federal agencies, immediately start the process to modify its rules to allow "general authorized access" devices to operate in two bands in the NTIA Fast track list, specifically the 3550-3650 MHz (radar) band and another to be determined by the NTIA and FCC. A feasible way to operate this system would be as an extension of the White Space system being developed and deployed by the FCC and various third party vendors in the TV Bands, but NTIA and FCC should determine the most appropriate management technology.
>
> The rules for this use will require the general authorized access devices to be both registered and frequency agile. Over time, these bands should also be migrated to the system for repurposing Federal spectrum outlined in the other recommendations of this report. The migration of these bands will be the most immediate item overseen by the White House Chief Technology Officer (CTO) in bringing together the Spectrum Management Team (SMT) created by recommendation 5.1.

A convergence of technology advances makes it possible to easily allow very limited but valuable general authorized access spectrum sharing in certain regions of Federal spectrum almost immediately. In the NTIA Fast Track List,[132] the 3550–3650 MHz band with radar systems is a good initial candidate.

A second band should be chosen by the NTIA. Some good candidates include the 1675–1710 MHz band, 406.1–420 MHz band, the 4950-5000 MHz band, or expanding the 3550–3650 MHz band with the adjacent 3650–3700 MHz band. It would be particularly useful to allow general authorized access to a band below 2.5 GHz, in addition to the 3550–3650 MHz band, that has propagation characteristics suitable for mobile broadband applications (in addition to fixed wireless). Spectrum with better propagation will encourage investment in spectrum sharing technologies and services useful in developing the market for small cell and other shared spectrum solutions. To accommodate sharing of "general authorized access" devices in these two bands without compromising the Federal services already using them, a white spaces database could be easily extended to accommodate device registration and spectrum sharing exclusion zones as determined by the NTIA.

Ten third party vendors[133] who are already developing or operating databases in conjunction with the FCC for commercial TV bands are immediate candidates to develop and operate databases for these added Federal bands.

A recent analysis of Federal spectrum reallocation options[134] provides the basis for establishing spectrum sharing exclusion zones in these two bands, which we refer to as the White Space(WS)–Federal addition. This analysis should be used to derive exclusion contours, similar to those used by the commercial White Space database process in the TV bands, making possible a wider range of candidate uses, particularly low power and Wi-Fi-like local area uses. Initial development of this analysis may have to be conservative to meet aggressive sharing schedules.

The extension of the White Spaces database governing access to Federal spectrum must also reflect the dynamics of Federal system operations, which will in some cases require near-real-time interfaces from Federal users to the database and, for classified systems, no more information than the interference limits governing access. Initially, these gating criteria can be very conservative, particularly in bands where DFS or other sensing or beaconing technologies will be needed as an additional or alternative safeguard to database governance.

The interface between the WS–Federal addition implementations and devices is one that should be developed by industrial suppliers of the WS service. While it would be desirable to have a standard for implementation of this interface, industry is in the best position to define this standard as providing it though a Federal regulatory proceeding would constrain the availability of this service quickly. Besides requiring device registration, the radios in these devices must also have the ability to operate in more than one frequency so they can continue to operate as these bands are repurposed to the long term Federal spectrum repurposing system and SAS outlined in this report.

Implementation of the WS–Federal addition should not impact the viability or prospects of any future relicensing actions, including reallocation of spectrum segments through auctions. The primary Federal users that are protected through the SAS mechanism would be replaced by the resulting civil user, and would have assured non-interference through the same mechanisms that protected Federal users.

The secondary devices that were present in the band would continue to operate in non-interfering conditions, and spectrum availability would be driven by the rate and coverage of the primary user build out. Similarly, implementation of the WS–Federal addition should not preclude migration of the two Federal bands to a long term, SAS-based system. In fact, implementation in these two Federal bands will allow us to observe real world dynamic sharing with Federal systems to better learn lessons when implementing the SAS system. As with White Spaces sharing, we believe the initial devices in these two bands should be low power, short range (high spatial reuse) devices.

These two bands could then also serve as the pilots to start measuring effectiveness and effectiveness improvements over time using the metric proposed in Section 2.2 of this report. This will, hopefully, give the NTIA and the Federal agencies a lot more knowledge in the planning cycles towards the new (big block) architecture proposed for repurposing Federal spectrum in Section 2.1. Additional costs to the Federal Government, beyond NTIA data collection as required for the Fast Track reports, should be minimal as most costs will be absorbed by industry. A comparison of the functionality of the White Spaces database, the WS–Federal addition, and the Spectrum Access System is provided in Table 7.1.

7.7. Costs and Funding

Many of the recommendations of this report ask for policy or organizational initiatives that will incur no or minimal direct costs. Three areas, however, imply some modest costs: data collection on spectrum usage by NTIA to help guide further spectrum planning around a new architecture and dynamic sharing; setting up and running the Spectrum Access System; and setting up and running the Test City and Mobile Test Service. In the second and third of these areas we expect that in the medium to long term, a significant portion of the costs will be taken on or shared by commercial entities.

NTIA Data Collection: We estimate a total cost of $2-3 million a year for an additional 10-15 people at the NTIA and Federal agencies to collect and improve the data required to improve the grouping allocations to move to a super highway architecture.

In out years, there will likely be further costs to procure an advanced allocation system, but the current staff at the NTIA should evolve to handle the new system, without additional headcount, by upgrading the inefficient way this work is done today. If legislation to evolve the Spectrum Relocation Fund into a Spectrum Efficiency Fund is enacted the next year, it could provide a firm basis to move some of these R&D needs forward.

Table 7.1. Functional Comparison of White Spaces Database, WS–Federal Addition, and Complete SAS

Functionality	White Spaces (WS in TV Bands)	WS–Federal addition	Spectrum Access System
Accept Specific Interference Contours for Federal Primary Access users and Specific Secondary Uses	✔	✔	✔
Automatically Determine Interference Possibilities for any Secondary Technology			✔
Register the Location of Secondary Devices Authorized to Operate		✔	✔
Provide Deconfliction of Secondary Spectrum Users			✔
Provide Real Time Input of Primary User Operating Locations and Periods		✔	✔
Provide Marketplace for Leasing of Spectrum and Revenue to Treasury			✔
Provide the SMT Metrics and advanced features like Time to Live (TTL)			✔

Spectrum Access System (SAS): We estimate a Federal cost of $10 million, divided among four categories:

- **Data collection from the Federal Systems:** Depending on the amount of data initially required and how much it could leverage the current NTIA systems, this could be just $10 million or less over the first three years, assuming we start with more static bands and set conservative limits. As more complicated bands are added, with a lot of dynamic characteristics, these costs may increase as dynamic data is required. We would hope those costs could be absorbed by an increment to the Test City funding sources, as the SAS would be used in the Test City, or through the Spectrum Relocation Fund, assuming part of the 1,000 MHz region would be offered in some geographies for initial revenue.
- **Building and deploying the SAS:** Ideally, this would work on the same model as the White Space databases for the TV Bands, for costs are absorbed by the private sector. White Spaces initially found 10 partners to each build systems. In the case of

SAS, we expect it to take 10-20 years of software development for each third party to have product ready for deployment.

- **Addition of Digital Rights Management (DRM) capability to existing and new Federal devices in order to provide security and reliability for transmission requests and permissions sent between devices and the SAS:** Initial provision of this capacity may be expensive, but over time, we expect it will become a standard component of newer radio systems. We expect any necessary funding to come from the usual agency equipment upgrade budgets or, over time, from the Spectrum Efficiency Fund.
- **Cost of commercial users to join in sharing use of Federal bands:** We expect any such costs to be absorbed by commercial users as part of their business model for participating in spectrum sharing, and therefore to pose no Federal budgetary concerns.

Test City and Mobile Test Service: We estimate that construction and the first three years of operation of these facilities will cost about $60 million. Initial funding, depending on timing and other factors, could come from the Wireless Innovation Fund (WIN)[135] that was part of the Middle Class Tax Relief and Job Creation Act of 2012,[136] passed in February 2012, as the testbeds were also planned to be used for Public Safety, which would be an integral part of the Test City.

The initial WIN fund, which is controlled by NIST, is $100 million (it adds another $200 million in later milestones). It may be possible to also entice the private sector to be part of the first three years of funding in either dollars or manpower and capital equipment in-kind. DARPA could also be a good funding source, especially for the Mobile Test Service, as the results of the tests will help DOD better understand how to modify their systems to accommodate sharing. After three years, we would hope that ongoing operations support would come from private companies executing tests in the Test City or Federal agency budgets testing systems against the Mobile Test Service.

In the longer term, we believe that the test facilities could operate as Public Private Partnerships (PPPs), which represent a versatile mechanism for Federal, state, and local governments to pool their resources with industry and academia to foster technological innovation among U.S.-owned firms and the Federal Government. Total Costs: Combining these estimates, we estimate that initial costs will be about $80 million for the first three years and that ongoing funding would be under $10 million a year once the test systems are self-sustaining through private sector funding and upgrades to Federal agency equipment are regularly budgeted through the Spectrum Efficiency Fund.

7.8. Timeline

Table 7.2 outlines the timeline for implementation of the recommendations of this PCAST report. The actions are grouped into immediate (within 1 year), near term (1 to 3 years), medium term (3 to 10 years), and long term (more than 10 years).

Table 7.2. Timeline for Recommendation Implementation

	Less than 1 year	1-3 years	3-10 years	10+ years
Recommendations				
1.1: Issue President memorandum on share 1,000 MHz of underutilized Federal spectrum	Issue Presidential Memorandum Identify possible Federal Spectrum for Sharing	Provide initial sharing access to underutilized Federal spectrum	Reduce need for Federal spectrum relocation by proving sharing model with receiver management	Fully implement Federal spectrum architecture
2.1 Establish a mechanism to manage the sharing of Federal spectrum 2.2: Authorize and implement, Federal Spectrum Access System (SAS)	Specify database and sensing technologies needed for Spectrum Access System and Secure initial third party partners	Implement Spectrum Access System with primary Federal users and initial "General Authorized Access"	Implement secondary access users into Spectrum Access System and provide revenue collection capability from exclusive spectrum access.	Fully implement multitier Spectrum Access Systemas basis for operation of shared Federal bands
3.1: Establish spectrum management methodologies that consider both transmitter and receiver characteristics	Develop receiver management framework	Implement receiver frame-work for new assignments in and adjacent to initial bands	Implement receiver framework for new and legacy Federal sharing assignments	Use receiver regulation models to increase effectiveness of sharing assignments
4.1: Base Federal spectrum policies on their innovation effects and market growth			Trial new economic licensing models (medium and short term) Collect initial revenue from new licenses	Collect significant revenue from new spectrumeconomic usage models
5.1: Formalize Spectrum Management Team (SMT) 5.2: Reexamine the partitioning of Federal spectrum 5.3: Establish mechanism to share Federallyheld spectrum 5.4: Implement a Federal agencies incentives program 5.5: Implement a sustainable funding mechanism 5.6: Appoint Spectrum Sharing Partnership Steering Committee (SSP) 5.7: Harmonize internationally and regional wireless broadband spectrum allocations	Create SMT Convene SSP Create Federal usage baseline using Spectrum Currency Create Rules to evolve Spectrum Efficiency Fund from Spectrum Relocation Fund Establish quantitative goals for Federal spectrum sharing	Implement Spectrum Currency incentives program with regular monitoring Implement Spectrum Efficiency Fund Begin international and regional discussions on harmonization of newly shared spectrum bands	Integrate Spectrum Currency with Spectrum Efficiency Fund and help fund new Federal efficiency projects	Create ongoing programs to monitor and fund the improvementof Federal spectrum usage and sharing

	Less than 1 year	1-3 years	3-10 years	10+ years
Recommendations				
6.1: Provide test services (a Test City and a related Mobile Test Service)	Create specifications for scalable Test Services Secure funding for testbeds, spectrum access system, NTIA data collection	Offer testbeds to private and public sector for experiments Use testbed results to imple-ment spectrum sharing technologies and criteria in commercial products	Use testing results to start to upgrade Federal equipment Secure ongoing funding via robust privatepublic partnership	Deploy new generations of Federal systems that inherently support Federal spectrum sharingtechnologies
7.1: Allow near term "general authorized access" devices to operate in two Federal bands	Complete FCC rulemaking (NPRM) Create conservative sharing maps Extend White Spaces with registration	Enable operation of "general authorized access" devices to operate in two Federal bands	Cutover initial operations in these bands to Spectrum Access Systems	

APPENDIX A: ACKNOWLEDGMENTS

PCAST expresses its gratitude to the following individuals who contributed to this report by attending meetings or responding to requests for information:

Ian Atkins
Group Manager Spectrum Engineering
Federal Aviation Administration

Byron Barker
Chief, Strategic Planning Division
Office of Spectrum Management
National Telecommunications and Information Administration

Robert J. Baxley
Senior Research Engineer
Georgia Tech Research Institute

Coleman Bazelon
Economist
The Brattle Group

Colin Blackman
Director
Camford Associates

Jeff Bratcher
Chief, NTIA/ITS.P
PSCR Technical Manager
Institute for Telecommunication Sciences
Department of Commerce

Dean Brenner
Vice President, Government Affairs
Qualcomm Inc.

Donald Brittingham
Vice President, National Security and Public Safety Policy
Verizon

Robert W. Brodersen
Co-Scientific Director
Berkeley Wireless Research Center

Mary Brown
Director of Government Affairs
Cisco Systems

Leonard J. Cali
Senior Vice President, Global Public Policy
AT&T Inc.

Vinton G. Cerf
Vice President and Chief Internet Evangelist
Google, Inc.

Vincent W.S. Chan
Joan and Irwin Jacobs Professor of Electrical
Engineering and Computer Science
Massachusetts Institute of Technology

John Chapin
Program Manager, Strategic Technology Office
Defense Advanced Research Projects Agency

Aneesh Chopra
United States Chief Technology Officer

Charles Clancy
Director, Hume Center for National Security
and Technology
Virginia Tech

David Clark
Senior Research Scientist
MIT Computer Science and Artificial
Intelligence Laboratory

Andrew Clegg
Program Director, Enhancing Access to the Radio
Spectrum (EARS)
National Science Foundation

Klaus Doppler
Principal Researcher Nokia
Research Center

Rashmi Doshi
Chief, Laboratory Division
Office of Engineering and Technology
Federal Communications Commission

Edward Drocella
Chief, Spectrum Engineering and Analysis Division,
Office of Spectrum Management
National Telecommunications and
Information Administration

Håkan Eriksson
Senior Vice President and Chief Technology Officer
Ericsson

Simon Forge
Project Manager and Director
SCF Associates Ltd.

Gardner Foster
Assistant Chief, International Bureau
Federal Communications Commission

Jason Furman
National Economic Council
Executive Office of the President

Kaigham (Ken) J. Gabriel
Deputy Director
Defense Advanced Research Projects Agency

Patrick Gallagher
Director
National Institute of Standards and Technology

Andrea Goldsmith
Professor of Electrical Engineering
Stanford University

Christopher Guttman-McCabe
Vice President, Regulatory Affairs
CTIA – The Wireless Association

S.M. Hasan
RF Engineer
GE Global Research

Robert Horvitz
Open Spectrum Alliance

Frank Jager
Distinguished Member of Technical Staff
Wireless Technology Strategy
Verizon

Walter Johnston
Chief, Electromagnetic Compatibility Division
Federal Communication Commission

Kevin Kahn
Intel Senior Fellow and Director of
Communications Technology Intel Labs

Rick Kaplan
Chief, Wireless Telecommunications Bureau
Federal Communications Commission

Dina Katabi
Associate Professor of Electrical Engineering and
Computer Science
Massachusetts Institute of Technology

Julius Knapp
Chief, Office of Engineering and Technology
Federal Communications Commission

Peter L. Levin
Senior Advisor to the Secretary and CTO
Department of Veteran's Affairs

Narayan B. Mandayam
Professor of Electrical and Computer Engineering
Rutgers University

Madhav Marathe
Deputy Director, Network Dynamics and
Simulation Science Laboratory
Virginia Tech

Michael Marcus
Director
Marcus Spectrum Solutions LLC

Joan Marsh
Vice President, Federal Regulatory Affairs
AT&T Inc.

Joseph Mitola
Vice President of Research Enterprise and
Distinguished Professor
Stevens Institute of Technology

Harlin R. McEwen
Chairman
Public Safety Spectrum Trust

Paul Mitchell
Technology Policy Group
Microsoft Corporation

Sumedha Mohan
Graduate Student, Department of Electrical and
Computer Engineering
Virginia Tech

Joseph Molnar
Section Head
Systems Integration and Instrumentation Section
Naval Research Laboratory

Karl Nebbia
Associate Administrator
Office of Spectrum Management
National Telecommunications and Information
Administration

James Neel
President and CEO
Cognitive Radio Technologies

Eric Nelson
Chief, Measurements Division
Institute for Telecommunications Sciences NTIA

Michael Nelson
Research Associate, Leading Edge Forum
Visiting Professor of Internet Studies
Georgetown University

Ali Niknejad
Associate Professor of Electrical Engineering and
Computer Sciences
University of California at Berkeley

Eli Noam
Professor of Economics and Finance
Columbia Business School

Dereck Orr
Program Manager
Law Enforcement Standards Office
National Institute of Standards and Technology

Janine Patterson
National Security Staff
Executive Office of the President

Jon M. Peha
Professor of Engineering and Public Policy,
and of Electrical and Computer Engineering
Carnegie Mellon University

Craig Peters
Senior Economist
Council of Economic Advisers

Danny Price
Director of Communications and Spectrum Policy,
ASD(NII)
Department of Defense

John Quinlan
Office of Management and Budget

Theodore S. Rappaport
William and Bettye Nowlin Professor of Engineering
University of Texas

Charla Rath
Vice President, Wireless Policy Development Verizon

Dan Reed
Corporate Vice President,
Technology Policy Group
Microsoft

Frank Sanders
Chief, Telecommunications Theory Division
National Telecommunications and Information Administration

Gregry Schaffer
Assistant Secretary for Cybersecurity and Communications
U.S. Department of Homeland Security

Robert Schneider
Chief of the Spectrum Enterprise Services
Division of the Defense Spectrum Organization
Department of Defense

Ivan Seskar
Associate Director of Information Technology
Wireless Information Networks Laboratory (WINLAB)
Rutgers, The State University of New Jersey

Tom Shaginaw
University Research Program Manager
General Electric Co.

Sanyogita Shamsunder
Director of Technology
Verizon Wireless

William H. Stone
Executive Director of Network Strategy Verizon

Lawrence E. Strickling
Assistant Secretary for Communications and Information,
Department of Commerce Administrator
National Telecommunications and
Information Administration

Rangam Subramanian
Principal, Wireless Technology & Business Strategy
Idaho National Laboratory

Joel Szabat
Deputy Assistant Secretary
Department of Transportation

Teri M Takai
Chief Information Officer
Department of Defense

Peter Tenhula
Vice President and General Counsel
Shared Spectrum

Stuart Timerman
Director, Defense Information Systems Agency
Defense Spectrum Organization

Steven VanRoekel
U.S. Chief Information Officer Executive
Office of the President

Hal Varian
Chief Economist Google, Inc.

Brett. T. Walkenhorst
Senior Research Engineer
Georgia Tech Research Institute

Patrick Welsh
Assistant Vice President,
Public Policy Verizon

Tony G. Werner
Executive Vice President
and Chief Technology
Officer
Comcast

Buckmaster de Wolf
Senior Counsel Litigation
and Legal Policy
General Electric Co.

APPENDIX B: PROPOSED NEW METRIC FOR SPECTRUM USE

B.1. A Metric for Scalable Wireless Broadband Networks

In Section 2.3, we argued that spectrum utilization needs to be measured with a more sophisticated metric than simply data transmission divided by bandwidth.

This appendix presents some examples of metrics that take into account not only how well a given user can transmit data, but also the extent to which that transmission precludes other users from using spectrum. More complex metrics are required to design operations systems, but this metric is sufficient to represent the significance and impact of the recommendations in this report. This metric does not addresses non-communications use of spectrum, but the bulk of the new applications for spectrum by non-Federal users are communications related.

The general form of the proposed metric is: [137]

$$\mathrm{Eff}_{Spectrum} = \sum_{n=1}^{k} \frac{R(n)D(n)}{I^2(n)T(n)S(n)} \tag{1}$$

Where:

EffSpectrum	Spectrum Effectiveness	Spectrum Effectiveness in terms of data delivered across a range, over the spectrum, area, and time whose usage is precluded
R(n)	Communication Range	User n's actual communication range
D(n)	Data Delivered	Quantity of data delivered for user n
I(n)	Interference Range	User n's interference range, out to which other uses of spectrum are precluded
T(n)	Transmitted Data	Quantity of data actually communicated to user n
S(n)	Spectrum Precluded	User n's actual spectrum precluded to other users
k	Number of Users	Total spectrum users within a block of spectrumand over a region of operation

For static systems such as some cellular downlinks, we can simplify the equation to the steady state by:

- replacing the time and data delivered term with a general expression of peak capacity, C_0
- considering the denied area to be constant and equal to $I_0{}^2$
- considering that users are distributed across a range of 0 to R_0, with a mean of kR_0 The effectiveness measure of Equation (1) becomes

$$\mathrm{Eff}_{Spectrum} = \frac{kR_0C_0}{I_0^2S_0} \tag{2}$$

A similar measure can be developed to consider the density effects of different system architectures. In this case, we ignore the range term, since we would be considering comparisons of alternative methods to accomplish identical communications tasks, for example by considering the effectiveness of different architectures to deliver service to the same set of nodes.

If the serviced set of nodes is equivalent in two architectures, the effectiveness of this architecture reduces Equation (2) further to:

$$\mathrm{Eff}_{Architecture} = \frac{C_0}{I_0^2 S_0} \tag{3}$$

where Eff $_{Architecture}$ is measured in terms of data capacity over the spectrum and the area whose usage was precluded.

This metric has units of *Bits/Area Hertz.* It can be used to compare, for example, the effectiveness of services provided by a cellular tower or by Wi-Fi to the same set of nodes, and helps explain why Wi-Fi is so effective that the small slice of spectrum it occupies was reported to offload over 40% of one carrier's smartphone traffic, despite its limited coverage and availability. Given the fundamental limits of communication theory, the only viable method to active density of information is to assure locality of communications.

This metric is important beyond just engineering applications or design of radios. Each of the terms in these equations is driven by policy, investment, and regulatory decisions, and all are therefore relevant when considering appropriate spectrum policies and investments. Most of the recommendations in this report relate to increasing the effective use of spectrum through the policy impact on one or more of these variables.

The impact of the *S* variable (spectrum precluded) is very small. While we seek to increase bandwidth many times over (the FCC Broadband Plan converges on 50 times), the available spectrum that could be allocated to broadband wireless is only several times what is in use today. Therefore, this report concludes that simply trying to increase spectrum allocation is not an effective or sustainable strategy. Any new spectrum would have to not only increase the availability of spectrum for wireless broadband, but also enable fundamental changes in the architecture by which wireless broadband was delivered.

The *R* variable (communications range) makes it apparent that significant increases in broadband can be achieved through use of shorter range architectures, such as those that have emerged in microcell and femtocell technology and in unlicensed Wi-Fi. Spectrum that might not be suitable for high power, tower-based architectures may be very effective when applied in lower power, much more local architectures. Spectrum that was considered unshareable when the application was LTE may be appropriate for exactly the short range architectures that are most suitable for exponential increases in user bandwidth.

The *I* (interference radius) term illustrates the importance of reducing the interference range of systems, and has implications for both the transmit and receive side of the systems. Reductions in the transmitted bits/Hertz reduce the interference footprint as a ratio of the communications range. Transmit waveforms should transition from maximizing the bits/Hertz in scarce spectrum to instead optimizing for spectrum reuse. This optimization again favors a move toward lower power architectures, consistent with the concept of spectrum sharing. On the receive side, increases in interference tolerance are as significant as increased spectrum. If systems are more interference tolerant, then the reduction in the interference exclusion zones can readily increase capacity equivalent to the 500 MHz proposed for wireless broadband through changes in the architecture of the receivers and signal processing.

Lastly, the implication of the definition of the spectrum precluded term (*S*) implies that spectrum policies and allocations must include not only the spectrum occupied by the transmitted signal, but also any spectrum that must be left "fallow" to protect other users. This is typically done through the provision of "guard bands" between spectrum allocations. Examples of the constraints that this leads to are the Nextel/public safety and LightSquared/GPS usage conflicts described in Section 3.2. Much of the present inefficiency in resolving such conflicts is due to the necessity to assume that adjoining users have the poorest possible receiver protection. Poor receivers are externalities, as they impose a spectrum usage cost on other users of the spectrum. Effective and flexible spectrum usage requires that all users be able to predict the performance of adjoining devices, and that devices have sufficient performance that they do not preclude other users of the spectrum.

B.2. Metric Quantification and Examples

The metric reflects several of the fundamental architectural principles of modern wireless systems. Even if all of the usable spectrum below 3 GHz was provided for cellular usage, it would only increase capacity by approximately a factor of 5, and would be inadequate to meet even a few years'growth in wireless broadband usage. It is anticipated that there will continue to be fundamental advances in waveform design, signal processing, antennas, and antenna processing, but these will be largely evolutionary and bounded by Shannon's limit. Therefore, it is likely that the major increase in wireless broadband will have to arise through fundamental architectural changes.

Although the metric represents the architectural drivers abstractly, some concrete examples of the major factors in aggregate capacity are useful to obtain insight into its application and the opportunities that can be exploited through spectrum policy.

Shorter Ranges: Initial cellular deployments were designed to ensure coverage using implied high-power stations, positioned as high in elevation as practical. However, in providing aggregate capacity across a large number of users, the critical factor is the number of access points, not the capacity of a single access point. The number of access points provided is itself largely a function of the degree to which spectrum is reused, and the required spacing between access points on the same frequency. The reduction of node spacing (and thus increase in density) from what is required in low frequency, high power cellular systems to the ranges typical in Wi-Fi is instructive. Wi-Fi channels are almost interference free with spacing in a range from 30 to 100 meters in typical environments. Compared to spacing of 1,500 meters for a high-powered cellular, this yields a potential architectural benefit of from 225 to 2500 times capacity, for equivalent access technology. Additionally, in dense, high-rise environments, spectrum is often reused vertically (for example, separate Wi-Fi access on individual floors), increasing spectrum density by another factor of ten or more times in suitable environments. Of course, these dense environments are not typical, but they are the environments that are most stressing for conventional architectures to serve a high number of users with high quality of service broadband.

These benefits are not hypothetical. One major carrier is offloading more than half of its smartphone traffic onto Wi-Fi. Since 2.4 GHz Wi-Fi spectrum is significantly less than the spectrum available to that carrier, and that carrier's Wi-Fi traffic is only a small portion of total Wi-Fi traffic, it is apparent that this architecture is both effective, and in the process of being adopted, at least opportunistically.

This is an important consideration from the perspective of sharing Federal spectrum. Spectrum that may not be clearable and nationally available may be very well suited to these more localized usage models. The opportunity provided by sharing Federal spectrum is very well matched to the needs of deploying scalable national wireless architectures.

Spectrum Abundance: Creating spectrum abundance not only provides more possible users, but also enables users to be more effective in using spectrum within their systems. A user with only a single, narrow channel has to transmit a large number of bits on each Hertz of spectrum. Shannon shows that linear increases in bits/Hertz require exponential increases in power, but higher power leads to a greatly increased interference radius for a given communications radius. For example, assuming that the proposed spectrum policies can create wider band opportunities, modulation could be reduced from a value of 8 bits per Hertz to one, and still maintain the same link throughput. In this case, the power could be reduced by a factor of 132. In a line of sight communications path, this equates to a reduction in range of approximately 11.5, and a possible increase in density of 132. The most effective spectrum policy is not to maximize the throughput of a small segments of spectrum, but is to provide wide extents of spectrum, and to reuse it extensively.

One of the recommendations of this report is to reverse the movement to partition spectrum into smaller, more fragmented segments. Offering users wide segments of spectrum over smaller areas (rather than small segments of spectrum over large areas) is more effective from both a spectrum and service perspective.

Interference Tolerance: Wi-Fi can operate effectively with a large number of access points in very close proximity. When designers assume clear spectrum, they design systems that require clear spectrum. When designers are given the challenge of designing for highly contended environments, they can design systems that are highly tolerant of interference conditions and can operate in close proximity to each other. An example of this is the unlicensed band, which can simultaneously support Bluetooth devices along with a large number of Wi-Fi access points. Current policy, however, attempts to provide essentially interference-free channels to all users. A statistical model that might have up to 6 dB of noise floor elevation to some users at some times would enable separation distances to be decreased by at least factor of 2, for another increase in aggregate capacity of at least 4 times.

B.3. Summary

The aggregate bandwidth possible through these mechanisms could range from 900 to as much as 1.3 million times more than a fixed, large cell based architecture can provide. Of course, no single implementation will achieve all of these benefits. Realistic constraints will preclude some options in some situations. However, it is clear that spectrum availability need not preclude wireless broadband capacity, as spectrum sharing and reuse provide mechanisms to scale architectures that are appropriate to the density of users, and their wireless needs.

The opportunity to share Federal spectrum provides a policy mechanism to enable U.S. enterprises to implement this scalable architecture, without disruption to current services or

Federal users. It moves from small spectrum allocations across large areas to larger spectrum allocations in very localized regions.

APPENDIX C: TOWARDS A NEW APPROACH TO SPECTRUM ALLOCATION

C.1. Problems with the Current Allocation Process

As this report makes clear, traditional spectrum management practices by no means maximize spectrum efficiency. In fact, although there is a general perception of spectrum scarcity, most spectrum capacity is not used. An assigned primary user may occupy a band, preventing any other user from gaining access, yet consume only a fraction of the potential spectrum capacity. Unique among natural resources owned by the public, spectrum capacity is infinitely renewable from second to second—that is, any spectrum vacated by one user is immediately available for any other user.

Measurements of actual spectrum use show that less than 20 percent of the capacity of the prime spectrum bands (below 3.7 GHz) is in use even in the most congested urban areas.[138] The FCC's Spectrum Policy Task Force Report recognized this opportunity ten years ago:

> "Preliminary data and general observations indicate that many portions of the radio spectrum are not in use for significant periods of time, and that spectrum use of these 'white spaces' (both temporal and geographic) can be increased significantly..."[139]

However, the administrative process for enabling access to underutilized spectrum is expensive, inefficient, and slow. It can take many years for commercial users to successfully petition for re-assignment of unused or underutilized FCC-held spectrum. Although secondary markets for leasing unused or underutilized commercial spectrum exist, they are cumbersome and lack transparency. There is no such market for access to underutilized Federal bands. Private sector firms can seek Special Temporary Authority (STA) from the NTIA to operate on fallow Federal spectrum in a discrete geographic area, but such STAs do not offer private investors the scale, certainty or continuity they need to justify the capital cost of deploying network infrastructure or devices.

Presently, the NTIA and FCC coordinate interference protections through a bureaucratic process wherever there is a shared band in the table of allocations (and, for that matter, when there are potential interactions across non-shared Federal and commercial allocations). The coordination process, however, can be very time consuming, highly uncertain, and technically conservative, which can inhibit non-Federal use of the band. Moreover, the engineering analysis supporting interference protection criteria sometimes specifies how Federal uses are to be protected (e.g., through command-and-control restrictions on various operating parameters) rather than what is to be protected (e.g., power density at a protected site). The existing bureaucratic process poses a risk to dynamic sharing of Federal spectrum.

C.2. A New Approach for Allocation

This report argues for a fundamental change in spectrum management: Exclusive frequency assignments should not be interpreted as a reason to preclude other productive uses of spectrum capacity in areas or at times where the primary use is dormant or where underutilized capacity can be shared.

The principle of spectrum sharing was espoused in the National Broadband Plan, which recommended that "[t]he FCC should spur further the development and deployment of opportunistic uses across more radio spectrum."[140] In its rationale for that recommendation, the FCC's Plan observed that "opportunistic" and "cognitive" radio technologies "could allow access to many different frequencies across the spectrum chart that may not be in use at a specific place and time and could do so without harming other users' operations or interests."[141] The availability of both geolocation databases operating almost in real time and cognitive radio capabilities (i.e., sensing, dynamic frequency selection), working separately or in combination, make it feasible that spectrum capacity not needed for a primary user should be made available for shared access by secondary and opportunistic users on a band-by-band basis, subject to conditions that are tailored to avoid harmful interference to licensed operations.

Any effort to improve the spectrum allocation system must be mindful of several objectives. First, since spectrum is a public resource, maximizing its overall economic and social value should be a priority for management of Federal spectrum bands in particular. To that end, available technologies and governance mechanisms should be leveraged to open up access to as much underutilized Federal spectrum capacity as possible, and as quickly as possible, to meet the future needs of both Federal users and the private sector.

Second, when Federal bands are opened for shared access, it is essential both to safeguard Federal systems, especially in bands with national security and public safety uses, and also to permit these Federal systems and uses to evolve without being hamstrung by commercial or consumer uses that cannot accommodate changes in the conditions governing secondary access.

Third, the administrative system for spectrum sharing should keep labor costs and legal requirements to a minimum. Permissions to transmit and enforcement of conditions governing access should make use of technical solutions that automate most of these processes through the use of a management database and, where useful, spectrum sensing capability. Establishing a standard nomenclature for spectrum allocation and use across all bands of spectrum will further reduce administrative loads and make access more "machine readable," so that devices can register and access spectrum in a routine way.

Fourth, the registration and management of secondary Access and general authorized access users of shared bands must be done in a way that ensures Federal Primary Access users'continued ability to access spectrum as and when they need. A primary user may not yet be utilizing its spectrum pending a deployment of base stations, for example, or may use the spectrum in a particular geographic area only occasionally. The spectrum sharing system must not allow other users to "squat" on idle spectrum and refuse to vacate when the primary license holder requires access. The goal is to replace administrative processes that are used to evict Secondary Access users with faster and more effective technical mechanisms that guarantee the primary user's rights.

It is important to acknowledge that the use of shared spectrum will also have some impact on the types of technology that are optimal for shared bands. An example is the method of duplexing used in most cellular systems today. Because of the specific Federal spectrum that will be available to share in different areas will be variable and unpredictable, it likely will be extremely difficult to use Frequency Division Duplexing (FDD), as is traditionally used by most Commercial Mobile Radio Services (CMRS) operators today. This is because FDD requires the use of two channels of spectrum, along with duplex filters in base stations and equipment. The frequencies and spacing between the channels will vary with geography, even if two channels could be assigned on an exclusive basis in a given area.

However, with the ability to gain access to spectrum on a secondary exclusive basis, use of cellular carrier technologies like LTE can be accommodated. It is likely that the bulk of the systems that would share spectrum with Federal users would have to operate in Time Division Duplexing (TDD) mode. Although not used extensively in the United States, TDD is supported by LTE and many of the available chip set implementations. For example, Clearwire has announced plans to build its new LTE network in Broadband Radio Service (BRS) spectrum using TDD, and Qualcomm newest LTE chipsets support TD-LTE. WiFi and Bluetooth use TDD exclusively.

Although this report is agnostic about the technology to be used in the spectrum sharing regime, the private sector implementations of the SAS could include services that would enhance operational effectiveness across different entities or spectrum users. For example, SAS implementations could provide services, such as specifying a common timing profile using protocols like IEEE 1588, or through GPS timing that is used widely today, so that TD-LTE base stations operated by different entities can synchronize timing. This would shrink the interference contours of a secondary authorization by potentially a significant amount if TDD operation is widely used in given geographic area.

Chapter II of this report outlines a Federal Spectrum Access System (SAS) that can provide a three-tier hierarchy of access to Federal spectrum. The next section offers more details on the principles and functions of a possible SAS. We also assume that the SAS will be tied directly to a Receiver Management Framework, like the one described in Chapter III and, in illustrative detail, in Appendix D.

C.3. A Spectrum Access System to Govern Shared Access to Federal Bands

The Database Component of the SAS Envisaged in Section 2.4 Would Enable a Number of Distinct Operational Functions

Channel Selection/Permission: Devices would be certified for secondary and opportunistic access only if they have the capability to connect to the database (via a different band or form of access) and thereby periodically seek renewed transmit permission and/or terms of use updates, as appropriate for each band for which the device is authorized. Devices (or networks controlling devices) would periodically obtain permission to transmit by a valid, cryptographically-secured authorization from the database, which should be extensible for future coordination enhancements (such as spectrum sensing and cooperative radio protocols). It will be essential for Federal users to immediately report any new deployments, canceled deployments, or other changes in their actual use of spectrum, not just in their assignment from NTIA. In the aggregate, this registration of primary operations—and the associated analysis of necessary interference protection, such as exclusion zones—gives the SAS automatic ability to be queried and either give or deny permission to secondary access and general authorized access users.

Devices could rely on sensing alone for transmit permission, depending on the incumbent system (as 802.11 devices using sensing and DFS do today in certain Federal radar bands at 5 GHz), yet still check periodically for updates to terms of use. Generally, client devices under control of a base station may need to be registered (for identification in case of interference), but should not need their own direct connection to the database unless some form of peer-to-peer mesh capability is beyond the control of the base station.

Enforcement: While a combination of database access and sensing/DFS is likely to yield the most flexible and intensive use of unoccupied spectrum in many bands, the requirement of periodic and automatic connection to the database will ensure that no devices or networks are operating on out-of-date terms of use or without the capacity to be denied access a particular band when necessary. For this reason, as noted above, consumer devices will generally need to be "connected" and multi-band. The length of the authorization period between required contacts with the database could vary depending on the band and the nature of the primary use. For example, a relatively stable band with fixed-site primary equipment (e.g. certain radar systems) could authorize operation solely by sensing for extended periods, particularly if the device was also capable of responding to a signal beacon (or cognitive radio sensing) forcing it to immediately vacate the frequency until it checks in for reauthorization or updated terms of use. Since even devices relying on sensing and DFS (and/or a signal beacon) for the channel permission function will be required periodically to "call home" to renew their authorization, compliance with the current firmware and terms of use updates can be authenticated and enforced. Based on technical analysis of Federal systems, in bands where it is necessary to ensure that a malfunctioning device causing interference can be located, unlicensed devices could be required to register or to incorporate the capability to receive and switch off a frequency immediately in response to a narrow-band signal beacon (such as during an emergency when a Federal primary or other public safety use requires preemption of a band).

Entrant Equipment Validation: As described further below, authorization for shared access on a secondary or opportunistic basis would generally require a periodic authentication that the device is both certified for operation on that particular band and has been updated with the current terms of use for the band. This requirement anticipates the use of a public key cryptographic system to secure communications between the database and device or party seeking access. Authentication of users and the database authorization should run both ways: Users sign requests to the database cryptographically, so that the database can validate the identity and access rights of the requestor, and the database signs responses so that user equipment can validate that the spectrum use has truly been authorized. This system of FCC certification, registration and authentication through the SAS also facilitates enforcement, since the identity of errant devices, the manufacturer and recent location(s) could all be readily determined from transactional data collected by the database manager(s).

The SAS Would Also Include a Number of Other Operational Features

Uniform Interface and Ease of Use: The management database for all Federal spectrum bands should be implemented as a single, centralized system or as a decentralized system with components distributed by agency or band. We expect that although information concerning classified systems and assignments would not be disclosed to the public, the NTIA would make generic "terms of use" for those bands available to the SAS database (e.g., exclusion zones and transmit power limits). From the user's perspective, however, the database should be coordinated to provide a single consistent interface for accessing all geolocation database information and channel selection permissions. By maintaining all spectrum use in a common registry, a uniform interface analogous to the Internet's Domain Naming System (DNS) could be established that would allow users to find suitable spectrum for their application, either on a secondary or an unlicensed basis. This ease of access would enable greater opportunistic use of spectrum. This would also enable the requirement that secondary and unlicensed users must be multiband and capable of dynamic frequency selection, so that if a Federal primary user precludes use of a band (whether during an emergency, or permanently due to a system change), there is minimal risk of "stranded devices" or consumers whose devices suddenly become obsolete. The database would also promote sharing by enabling potential users to discover, in a straightforward way, areas where a given band of spectrum may be idle. Moreover, regulators or other users could examine the database to verify that spectrum is indeed being used in accord with listed registrations.

Time to Live Entries: The database should have "time to live" (TTL) entries, similar to the DNS, so that registrations and reservations for use are time-limited and renewed as appropriate. This is particularly important for Secondary Access users, who typically will be commercial entities receiving reserved or priority use on a Federal band in return for payment (see Chapter VII). These secondary leases (or licenses) to commercial users could take a variety of forms, including the auction of exclusive secondary use for extended periods in a particular area, as well as pay-for-use arrangements with relatively short TTLs. A user seeking to register a base station or transmitter in the database would request a period of time for which that registration will be valid. Any such request might be granted or denied based on the level of authority that the user has over a given band of spectrum.

For example, in a particular band the primary Federal user might request a TTL of one to five years, but a secondary user might be authorized only for 60 days at a time. A general authorized access user would most likely be denied any TTL, or at most a short pre-determined TTL compatible with the primary and anticipated secondary uses. Federal Primary Access users would then be assured of waiting only a limited amount of time to gain access or change terms of use as secondary authorizations expire. Conversely, by registering a short time ahead of the activation of a base station, the primary user provides notice of deployment to the database, so that existing Secondary Access users would know they would not be able to renew their allocation.

Security: Security is essential to the operation of the database, and it must run in both directions. The database must be able to determine that use requests are coming from authorized users, and users must be assured that the use authorizations they receive are valid. The system proposed here anticipates the use of a public key cryptographic system to secure communications between the database and device or party seeking access. Users would sign requests to the database cryptographically, so that the data-base can validate the identity and access rights of the requestor, ensuring that the requesting device is both certified for operation on that particular band and has been updated with the current terms of use, if required. Similarly, the database would cryptographically sign responses to requests, so that a device would only transmit in a given band after being assured of the validity of the authorization. This mechanism is conceptually similar to the technology for digital rights management (DRM) in securing media. In this way, even mass market consumer electronics devices can become registered Secondary Access users of spectrum, and primary license holders can be assured that these devices could not be transmit without valid authorization. Once a registration expires for a particular frequency band, the equipment will not be able to provide a valid key to transmit, although the equipment might be able to continue operating on other available bands, or to obtain secondary authorization later if the primary user's use was temporary. A number of unbroken DRM systems exist today, and adapting one of them to this kind of use would be straightforward if designing a completely new system was a problem.

For certain sensitive Federal systems, the raw user data needed to generate permissions for shared access (e.g., user identities, precise locations, duty cycles, frequency ranges) would not reside in a publicly accessible database, but rather in a linked and classified database secured within NTIA or elsewhere in the government. As a secured element of the overall SAS, this secure database would continually update any FCC or FCC-certified commercial database with information concerning the availability and terms of access to those particular frequency bands.

This system of certification, registration, and authentication through the geolocation database would also facilitate enforcement, since the identity, manufacturer, and recent location(s) of errant devices could all be readily determined from transactional data collected by the database.

Protecting the Use and Evolution of Primary Federal Users

The SAS includes many features intended to assure Federal agencies operating incumbent primary and secondary systems that shared spectrum access will be governed by band-specific conditions or terms of use that will safeguard their operations against harmful interference. Some of the features can be strengthened, if necessary. For example, the TTL mechanism could be extended to allow the database to contact the secondary device and immediately rescind its authorization to use the spectrum, shutting off the secondary user's access to the spectrum before their TTL expires. This option may be useful if a primary user requires the ability to preempt private sector use of a band during an emergency, detects interference exceeding prescribed levels, or can only give shorter than usual notice of a need to re-establish primary control of the spectrum in a given area.

It also important that safeguards for Federal Primary Access users can be updated as Federal users change their own operating parameters or if unforeseen interference scenarios emerge. As with commercial services and equipment, technological change as well as evolution of service needs will cause the spectrum requirements of Federal systems to evolve. Primary Federal users, particularly those related to national security or public health and safety, need to be sure that the terms of use for commercial secondary access to a band do not protect the secondary user's operational expectations or sunk capital investment in ways that compromise or preclude the primary user's ability to change its operating parameters. The SAS proposed here allays that concern by taking as a core principle that opening under-utilized bands for opportunistic access need not be permanent, or even long-term. Rather, both secondary and opportunistic users must be multi-band and capable of switching automatically to alternative frequencies if, in the future, authorization to one particular band expires or is denied.

The contingent nature of permission to transmit on an opportunistic or unlicensed basis would be similar to the FCC's White Spaces rules, whereby the Commission reserves the option to license additional TV stations, thereby removing a previously vacant channel from the White Spaces for TV Bands Databases in a particular local market, or to change a protected contour in reaction to evidence of harmful interference to a licensed entity. At any time, a band can be added, or withdrawn, or limited to use in a particular geographic area or at a particular time of day. Unlike in exclusively licensed bands, where it is expensive and time-consuming to upgrade or clear existing users, white space devices are not tied to a particular frequency. Bands can be opened or closed for sharing—nationally, regionally, or locally—and even on short notice, without stranding any users, legacy devices, or infrastructure.

An additional point of assurance for Federal users is that the database management system will increase the ability to evolve their services by enabling them to gain access to additional spectrum on a secondary or as-needed basis, and also by reducing, over time, pressure to relocate Federal users.

For spectrum sharing between commercial and Federal users to be effective, it will be essential to establish clear, generally applicable "rules of the road" governing the interference environment. In order to protect important Federal missions, NTIA and FCC should revise their coordination process to (1) ensure timely implementation of sharing frameworks, (2) focus on creating good "spectrum fences" rather than dictating "input" parameters, and (3) testing and trusting sharing technology, with appropriate safeguards, to avoid overly restrictive limits on spectrum use. Appropriately revised, the NTIA/FCC coordination process could provide a platform for increased spectrum sharing, rather than a barrier to it.

Sensing and Dynamic Frequency Selection

Although we envision that the database will be the core technical element by which the SAS manages spectrum sharing, a separate, complementary set of technologies can accurately sense and adjust to the actual radiofrequency environment in compliance with pre-programmed "policies" designed both to protect Federal Primary Access users and to deliver uninterrupted connectivity across multiple spectrum bands. The relevant concepts are variously referred to as "dynamic spectrum access," "cognitive radio," and "policy radios." These names are largely interchangeable and "basically describe radios and radio networks that can react and self-adjust to local changes in spectrum use or environmental conditions, to obtain access to spectrum without causing harmful interference."[142]

The FCC's Spectrum Policy Task Force recognized as long ago as 2002 that the cognitive radio technologies already being tested at DARPA and other labs would greatly enhance "opportunistic" access to unused spectrum:

> "Often technologies such as software-defined radio are called 'smart' or 'opportunistic' technologies because, due to their operational flexibility, software-defined radios can search the radio spectrum, sense the environment, and operate in spectrum not in use by others... That is, because their operations are so agile and can be changed nearly instantaneously, they can operate for short periods of time in unused spectrum."[143]

A basic feature of cognitive or policy-based DSA devices is that they operate only in accordance with prescribed policy constraints, which can be specified on a band-by-band basis. Moreover, the controlling software can be updated long after the devices are sold to end users. Conditions governing access to certain bands can be soft-wired into the device itself as well as into a geolocation database —and both can be regularly updated. [144]

For example, opportunistically sharing military radar bands is technically very different than sharing a band used primarily for fixed services, such as satellite or point-to-point microwave links, or a trunked land mobile radio system. One reason for DOD's willingness to allow dynamic sharing of radar frequencies in the 5 GHz band is that unlike television reception, for example, radar poses no "hidden node" challenge to spectrum sensing and DFS technologies, because the transmitter and receiver are co-located. In a fixed service band, by contrast, sensing may be less reliable than simply calculating the availability of frequencies in discrete locations from the database listing of protected transmit sites.

APPENDIX D: BETTER SHARING THROUGH RECEIVER REGULATION

D.1. Overview

As explained in Chapter III, regulators have traditionally put the onus on transmitters to limit adjacent channel interference by limiting their transmit power, or by mandating wider guard bands between transmitters and receivers. While no quantitative economic analyses of the trade-off seem to have been done, it is plausible that this was the optimal configuration in an era of few transmitters and a multitude of receivers, and a time when radio frequencies were easier to come by than cheap receiver filters. Nowadays, however, available frequencies are scarce, filters are cheap, and most devices are both transmitters and receivers. The need now to pack systems spectrally close to each other shifts the economic balance, and makes it essential for regulators and system designers to make reasoned trade-offs between receiver performance and transmitter permissions; in other words, to think about the radio system as a whole. This appendix goes into more detail about the nature of interference and the regulatory tools that can be used to manage the different classes of receivers, focusing primarily on managing the overload interference problem. There are further areas for review and improvements, but even small progress in overload interference is a big step forward. We also assume that any Receiver Management Framework will be integrated closely with a Spectrum Access System as described in Chapter II and in more detail in Appendix C.

D.2. Receiver Interference: A Brief Primer

Interference comes in two flavors: between operations using the same frequency range but in adjacent geographical areas or time slots, and operations using adjacent frequency bands in the same area and at the same time. The first type, referred to as co-channel interference, is managed by controlling the distance (in space or time) between systems using the same frequencies; the second, non-cochannel interference, depends on the frequency profile of the transmitted energy and performance of the receiver (see Figure D.1). Market forces typically work well to mitigate co-channel interference, as the users of the same frequency usually offer the same kind of service, and so have a shared incentive to deploy systems that can co-exist. We focus on the second case, non-cochannel interference Broadly speaking, interference across frequency boundaries can be due either to energy from a frequency neighbor falling in a receiver's assigned frequencies (due to imperfect filtering of the transmitter), or to energy outside the receiver's assigned frequencies that its receiver cannot ignore (due to imperfect filtering in the receiver). The degradation or interruption of a radio that results is a function of the entire system design comprising transmitters and receivers.

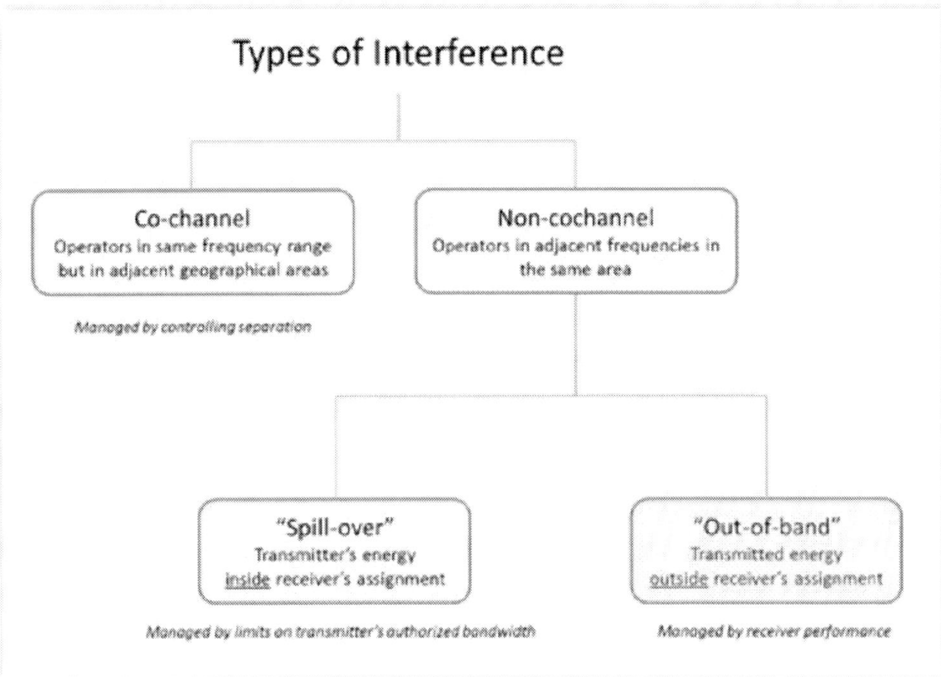

Source: PCAST.

Figure D.1. Interference Categories.

The first mode ("spill-over") is usually regulated via limits on emissions outside the transmitter's authorized bandwidth. The second mode ("out-of-band") involves several possible undesired responses of the receiver to the fundamental emissions in the transmitter's tuned channel. Since a receiver cannot filter out all energy outside its designated channel or band, it will pick up some energy transmitted by its neighbor, even if the adjacent transmission had a perfect filter. There are many mechanisms for this second mode of non-cochannel interference. For example, energy from outside a receiver's operating frequencies may desensitize a receiver, hiding desired signals, or it may generate signals within the operating frequency range by non-linear mixing in the receiver (known as intermodulation interference). The NTIA report on receiver standards gives the following list (NTIA 2003):[145]

- feed through of non-cochannel signals to the demodulator due to inadequate selectivity (filtering) at RF and IF stages;
- blocking due to an undesired very strong signal saturating the first amplifier stages and causing severe distortion;

- receiver desensitization resulting from erroneous automatic gain control responses to noncochannel signals;
- gain compression due to inadequate RF selectivity and dynamic range;
- spurious responses (to non-cochannel signals that mix with locally generated signals and fall within the receiver passband); and
- intermodulation of the desired and non-cochannel signals or two or more non-cochannel signals in non-linear stages of a receiver (e.g., in connection with gain compression).

A receiver's ability to process the desired signal in a frequency channel without being affected by interfering signals present in adjacent and other channels is described as its selectivity, and is largely determined by the following factors (see NTIA 2003 and Ofcom/TTP 2010[146]):

- Receiver channel filter performance (the ability to be able to receive large and small signals simultaneously): In modern digital receivers, this is divided into analog filtering prior to analogto-digital conversion (ADC) of the received signal, and digital filtering following the ADC. The quality of digital filtering is dictated by the dynamic range of the ADC.
- Reciprocal mixing: When the received RF signal is mixed with a local oscillator to convert it to typically a much lower frequency for channel filtering and demodulation, noise is added by the local oscillator that can swamp small wanted received signals.
- Receiver linearity: Nonlinearities in analog receiver elements of a such as amplifiers, mixers, and active filters introduce distortion to both the wanted and any unwanted signals that can lead to the creation of interfering signals at new frequencies. If these occur at critical frequencies within the receiver, they will affect the receiver's ability to receive a small wanted signal.
- Spurious responses: Unwanted signals at certain receiver-dependent frequencies could block the wanted signal.
- A receiver's behavior in the presence of such effects is characterized by a variety of parameters, including (NTIA 2003, section 2):
- Adjacent Channel Rejection (attenuation): The ability of a receiver to reject signals in the adjacent channel.
- Adjacent Channel Selectivity: The ability of a receiver to discriminate between a desired signal and an undesired signal in an adjacent channel.
- Image Frequency Rejection: The ability of a receiver to reject signals at the image frequency.
- Intermodulation Rejection (aka Cross Modulation Rejection): The ability of a receiver to reject intermodulation products produced by the mixing of two or more signals at the input to the receiver.
- Selectivity: Rejection (attenuation) of an undesired signal at frequencies close to the desired signal frequency. It is often specified as the amount of frequency difference between desired and undesired signals needed to produce a specified attenuation of the undesired signal.

- Sensitivity Depression or Desensitization: The level of a non-cochannel signal that increases a receiver signal power threshold or decreases receiver gain by a defined amount.
- Spurious Response (aka Spurious Rejection): Undesired receiver response resulting from mixing of the local oscillator and undesired signals. This includes the response to undesired signals at the image frequency.

We recommend that the technical implementation details of how a receiver manages to achieve a certain selectivity or sensitivity are left to the designer. In order to understand the impact that receivers have on the use of adjacent spectrum bands it is essential to understand the receiver's performance, but not how it is implemented.

Section 3.2 explains how receivers can be divided into five classes of receivers, based on the frequency assignments (Federal or non-Federal) in the band, and whether the receiver is under control of the spectrum licensee or not. For convenience, Table 3.1 is reproduced here as Table D.1.

D.3. Regulatory Tools

The rapidly growing use of spectrum increases the incidence of non-cochannel interference. Furthermore, the ability to share a given band will be limited by the capabilities of the receivers deployed in the band. It is therefore clear that more guidance from the regulator regarding receiver operation is required in order to optimize radio systems as a whole, seen as a combination of receivers and transmitters. Receiver interference limits are the first in a graded set of steps of increasing specificity, each building on the previous one (see Figure D.2), that the regulator can put into practice. Receiver interference limits define the interference the receiver is expected to tolerate; subsequent steps address device performance.

- Receiver interference limits
- Self-certification of device performance
- Mandated receiver standards

Table D.1. Five Classes of Receivers

Band Occupancy Class	Receiver Type
Only Federal assignments	All
Only nonFederal (FCC) assignments	Licensed
	Decoupled receivers in licensed service;Part 15 communications devices
NonFederal (FCC) assignments share band with Federal users	Licensed
	Decoupled receivers in licensed service;Part 15 communications devices

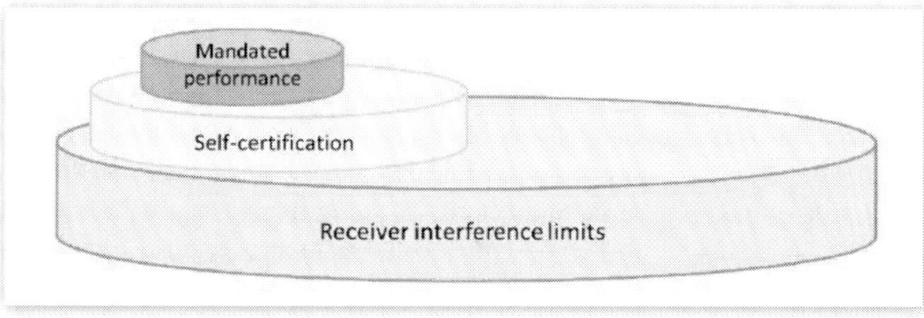

Source: J.P. de Vries.

Figure D.2. Progression of increasingly specific regulatory steps to manage reception.

Receiver Interference Limits

Delineating the radio interference that receivers should expect to operate in without being able to make claims of harmful interference is the necessary minimum; we call this receiver interference limits. The interference limit would be defined as a profile of field strength or power flux spectral density (e.g. in units of dB (W/m^2) per MHz or dB (pV/m) per MHz) over frequency, both in-band and out-of-band, at all locations in the licensed operating area, to be observed for a minimum percentage of times and places, with a specified measurement resolution in space and time. Note that the receiver interference limit does not specify or regulate receiver performance in any way; it is a definition of the interference the receiver is expected to tolerate. A receiver operator could only make a claim for harmful interference if the aggregate signal strengths from neighbors exceeded the interference ceiling. Receiver interference limits and transmission permissions would be chosen to be mutually consistent: that is, the aggregate energy permitted by authorized transmissions would be less than the designated receiver interference limit. The interference limits are defined probabilistically to take into account local and transient fluctuations of energy resulting from transmissions that would exceed any given ceiling. An interference limit is not an attempt by the regulator to define the interference environment, now or in the future. It is not a description of the environment, but rather a criterion for making a harmful interference claim that tries to make rights more objective. If there is already an incumbent in an adjacent band over which the interference limit is to be defined, the limit will be an upper bound to the environment that neighbor creates—most likely a very generous one for reasons of political and engineering prudence. Even though interference limits do not define the RF environment, they give radio designers more guidance on the environment than they currently have, which is at most the allowed transmit power of allowed operation in neighboring bands, with no information on the density/deployment of those transmitters. The interference limit gives resulting field strength at the receiver, which is what really matters in terms of performance, not power-at-the-tower as in Effective Isotropic Radiated Power (EIRP) transmitter rules. Given this information, the designer can then make the trade-offs between minimum desired signal strength (sensitivity), out-of-band filter performance and carrier-to-noise ratios required to deliver the specified quality of service for their system as a whole, that is, the combination of transmitters and receivers. In fact, interference limits allow designers to consider the overall system design; a communication system, say, can be engineered to operate successfully either by improving the adjacent channel rejection of receivers, or by increasing the strength of desired signal relative to the interference by reducing the distance between the receiver and transmitter (e.g. by increasing the number of transmitter locations) or through higher transmit power. While receiver interference limits and interference temperature[147] use similar units, they are used differently and have different goals in mind. One key distinction is that interference limits focus on solving out-of-band, cross-allocation interference, whereas interference temperature is concerned with in-band, co-channel operation. Further, interference limits do not grant second party rights in a primary licensee's frequency block, whereas interference temperature was designed to facilitate and encourage second party, co-channel operation. While interference limits will facilitate sharing by specifying the interference that devices will have to tolerate, that's in the context of adjacent channels, not co-channel. Thus it is only necessary to actually measure the interference, and compare to the protection limit, when harmful interference from adjacent channels is suspected.

In a sharing scenario, a device wishing to operate on a secondary or unlicensed basis is given an interference limit profile that it would need to be able to tolerate. This limit would be at least as high as the aggregate transmissions of the primary users. The secondary or unlicensed service would then have to determine whether it could operate satisfactorily given interference at this level. Receiver interference limits would be set by the regulator responsible for making the operating assignment, and would take into account the characteristics of current allocations and devices using them. For example, the interference limits for a radiolocation service would reflect the in-band and out-of-band interference that a radar receiver can tolerate; the allowed interference would not exceed these limits. The receiver interference limits for a communications service adjacent to radar would be set so that the communications licensee would have to tolerate the energy delivered by the radar transmitter in, and adjacent to, its licensed frequencies.

For example, if receiver interference limits had been set for the AWS-1 F block auctioned in 2008, it would have been clear at the outset whether handsets in the upper F block needed to accommodate TDD transmissions in the adjacent AWS-3 block, or not. Since unlicensed operations must accept interference from other transmitters (47 C.F.R. 15.5 (b)), their interference limit is effectively zero. However, we believe it is worthwhile setting non-zero receiver interference limits for unlicensed operations as a way of defining explicit minimum interference tolerance,[148] and coordinating operation among unlicensed devices. The translation of interference limits for licensees into transmission permissions for unlicensed transmitters in adjacent allocations can be done by analyzing likely deployment scenarios of unlicensed devices, using that to calculate the probability distribution of resulting aggregate signal strength, and then setting the transmission power for individual devices in a way that this result remains below the adjacent licensee's protection limit. If deployment densities of unlicensed devices are much lower than anticipated in the analysis, a rule change could increase their allowed transmit power; if the densities are higher, or evidence emerges of interference problems, it would be decreased. Experience shows that the FCC will err on the side of caution on setting these limits. In cases where a database controls unlicensed devices, this calculation and rule change could be done pretty quickly.

Additional Measures Beyond Receiver Interference Limits

The specification of an interference limit will be sufficient to incentivize efficient receiver operation in most cases. However, it may sometimes be necessary to augment it with device performance requirements, for example when there is reason to doubt that all devices that are built and deployed will be designed to operate satisfactorily given the interference limits. Moving beyond receiver interference limits requires the definition, by either the manufacturer or the regulator, of acceptable device performance in the presence of interference up to the protection limit. The adoption of receiver performance specifications in particular cases will require the examination of the value of such specifications; where they should be applied; who should participate; models for adoption; and policies. Options include a strong regulatory model (like the FAA); one where the FCC defers to industry but require some conformity tests; or nothing beyond interference limits. Developing device performance specification requires assumptions about quality of service parameters relevant to the use scenario, including the deployment strategy and licensee's tolerance of service degradation; and the definition of device performance parameter values that indicate that this quality of service is being met in the presence of interference up to the protection limit. The first option in specifying device performance would be for the manufacturer to self-certify that a device would be fit for purpose in its envisaged use, e.g. suffer no harmful interference in accordance with language of 47 C.F.R. § 2.1. Self-certification options include a retail warranty of fitness for purpose, conformance to an industry standard,[149] or the manufacturer submitting to the appropriate regulator a testing protocol and associated success criteria that allows testing of its claim of no harmful interference, e.g. following the FCC's equipment authorization protocol or the procedures in the NTIA Redbook.[150,151] In the case of the FCC, it might reserve the right to establish device performance standards if device suppliers fail to do so.[152]

The second option in specifying device performance would be for the regulator to issue performance mandates. It may choose to incorporate industry standards in rules, as is done by the FAA for aviation receiver standards,[153] or in the EU.[154] The regulator may choose to develop and promulgate its own device performance requirements, with the appropriate involvements of all stakeholders. These requirements would be promulgated in the NTIA Redbook or FCC regulations. Minimal receiver performance criteria such as a requirement that receivers implement a front-end filter with specified out-of-band attenuation may be sufficient. The most prescriptive approach would be for the regulator to determine what harmful interference means (e.g. quality of service criteria like acceptable bit error rate) and define device performance parameters (e.g. adjacent channel selectivity, EMC tolerance, image rejection, intermodulation rejection, and spurious rejection) that have to be met in order to ensure that a device does not suffer harmful interference as long as the radio interference is below the protection limit.

Enforcement

There is no requirement for the FCC to test all receivers. In cases where only receiver interference limits are given, assignments would not include any requirements on receiver performance, and no FCC equipment authorization would be required. Where self-certification is a warranty-of-fitness, i.e. the vendor says the device works to its intended purpose given the interference limit, non-compliance with this warranty would be via a false advertising claim. Where self-certification requires that the manufacturer submits a testing protocol to the appropriate regulator, the vendor would specify the quality of service metric that would have to be met in the presence of given desired signal and out-of-band interference at the interference limit. Assuming that the FCC approved the proposed criteria and testing regime, devices would be authorized following approval by the relevant testing lab. The case may arise where a receiver does not work satisfactorily and the manufacturer/operator blames the interference environment; that is, they make a claim of harmful interference against neighboring transmitters in adjacent bands. The plaintiff would bear the burden of proving that interference exceeds their protection limit, and that this is due to the transmitters. This could be done via field measurement or propagation modeling. Neither method provides absolute certainty: the number of possible field measurements is finite, and thus always merely a sample of reality as a whole; and the results of modeling depend on assumptions about propagation and terrain that cannot perfectly represent reality. However, perfection is not required (let alone feasible or desirable) in this or any other rights enforcement regime, just a well-defined arbitration mechanism that leads to a tolerably prompt, fair and certain outcome. It is advisable that the regulator specify the arbitration mechanism up-front for a particular allocation, e.g. a field testing protocol that specifies the spacing and time resolution of measurement locations, or a propagation model and a terrain data set.

D.4. Receiver Management Framework

We now combine the receiver management classes and regulatory tools to outline when particular receiver management regimes might be required. The first step, receiver interference protection limits, is recommended in all cases where transmission permissions are specified. Together, existing transmission permissions and receiver interference limits give designers guidance about the RF interference that devices are expected to tolerate. The other means are optional and would only be used in particular cases, as summarized in Table D.2. The degree of regulatory specificity will vary depending on context. In general, mandated device performance requirements will be rare, e.g. where there is a high likelihood that devices will be deployed that do not operate satisfactorily given the receiver interference limits leading to either interference susceptibility that cannot be remedied after the fact, and/or and risks to safety of life. This is most likely to occur with unlicensed devices, and receivers that are not operated or controlled by transmitter licensees.

- **For non-Federal use in all bands,** receiver interference limits would be set in all cases. Self-certification would be required only where necessary, e.g. to facilitate efficient coexistence among diverse systems. Mandated performance standards would be set only where essential, e.g. to protect safety of life in aviation systems.
- **For non-Federal use in FCC bands,** licensed operation would only be subject to the addition of receiver interference limits to operating rules. Unlicensed devices and decoupled receivers would be subject to device performance requirements, as follows. Self-certification would be required where it was likely that interference-intolerant devices will be deployed and later preclude other uses, e.g. a manufacturer's device volume exceeded 1,000 units a year. Mandated performance standards, e.g. the specification of a front-end filter performance specification, are a more interventionist step, and should only be used when it is highly likely that interference-intolerant devices will be deployed, e.g. a device manufacturer's volume would exceed 10,000 units per year.
- **For non-Federal use in bands shared with Federal users,** non-Federal licensees would be required to self-certify that they could operate satisfactorily in the presence of their designated receiver interference limits. Self-certification would also be required of all decoupled receivers and unlicensed devices, not just those that are

likely to be interference-intolerant, as is the case in bands not shared with Federal users. The requirement of mandated performance standards is also extended to cover cases where it is likely, not merely highly likely, that interference-intolerant devices will be deployed, e.g. a device manufacturer's volume would exceed 1,000 units per year.

Table D.2. Receiver Management Framework

Table D.2: Receiver Management Framework				
		Increasing regulatory intervention →		
	Regulatory Tools →	**Interference tolerance**	**Device Performance**	
Receiver Management Classes ↓		**Receiver Interference Limits**	**Self-certification Options include warranty-of-fitness, self-certification, conformance to industry standard**	**Mandated performance Options include front-end selectivity, mandated industry standards***
Federal assignments in all bands		all assignments	**If necessary**, e.g. to facilitate efficient coexistence among diverse systems	**If essential**, e.g. for safety of life in aviation
FCC assignments in non-Federal (FCC) bands	Licensed	all assignments	**Not necessary if** next to busy band that matches their interference limits **Required if** next to quiet band that's planned to be filled later	**If essential**, e.g. for safety of life in aviation
	Decoupled receivers in licensed service; Part 15 communications devices	all assignments	**Required if** next to a busy band that matches their interference limits	**Required if** next to quiet band that's planned to be filled later
FCC assignments in bands shared with Federal users	Licensed	all assignments	**Not necessary if** in a busy band that matches their interference limits **Required if** in a quiet band that's planned to be filled later	**If essential**, e.g. for safety of life in aviation
	Decoupled receivers in licensed service; Part 15 communications devices	all assignments	all assignments	**Required if** next to quiet band that's planned to be filled later

* Regulators should avoid imposing detailed performance requirements. Where such requirements are unavoidable, they should use voluntary consensus standards in lieu of government-unique standards except where inconsistent with law or otherwise impractical (cf. OMB Circular A-119 www.whitehouse.gov/omb/ circulars_a119#1).

The regulator should reserve the right to revisit and revise either the radio interference tolerance described by the interference limit, and/or device performance requirements. We recommend that this be done at fixed, predictable intervals.

For example, changes to license conditions such as interference limits would be notified during the license period, but only enacted upon renewal so as to provide appropriate time for the licensee to adapt.[155]

Moreover, the new regime can be implemented piecemeal, i.e. band by band. Each allocation is likely to have a different protection limit. Once a limit has been set by the regulator, it can be adjusted through bilateral negotiation between neighboring licensees. The following section describes in more detail how a regulatory regime can be put into effect and evolved over time.

Finally, we note that from time to time, wireless users acquire rights by dint of uncontested operation. This creates difficulties when interference arises with prior rights holders. The dispute between Nextel and public safety, and the operation of wireless microphones in the UHF band, have been characterized as instances of adverse possession, also known as squatter's rights.[156]

Since receivers operating on the assumption of quiet bands can preclude future planned assignments due to their adverse possession of these bands, it may be necessary to take steps upfront to avoid the squatter's rights problem. This issue is particularly pressing where sharing between Federal and non-Federal operations is planned. We recommend that assignments in bands adjacent to relatively quiet bands be given special consideration and require at least self-certification of devices deployed in these situations.

D.5. Implementation

Due to the differences in legacy usage, as well as the different objectives for Federal government and non-Federal commercial spectrum use, receiver management must be addressed differently by the NTIA and FCC. We start with receiver management in the Federal Government space, and conclude with a discussion of receiver management in non-Federal spectrum.

Federal Assignments

There is already considerable sharing between Federal users today, both in Federal-only bands and in bands shared with non-Federal users.[157] In the 1755–1850 MHz band, for example, 19 agencies operate 10 types of systems, with 3,183 individual assignments.[158] The sharing is enabled by defining unique operating regions along the traditional dimensions of space, time and frequency.

In order to expand usage by enabling more dynamic sharing methods, it will be necessary to carefully define and manage the receiver operations in these bands and at the adjacent edges of the allocations.

In order to facilitate more intensive and efficient sharing among Federal users, the NTIA should set and publish receiver interference limits using a transparent process for government assignments.

This will facilitate the deployment of automated assignment tools.[159] The addition of receiver interference limits to the other parameters in operating assignments can be rolled out in stages, starting with bands where intensive sharing is most likely and/or in bands where all the operations are under the control of single agency or department, thus simplifying administration.

Initially the limits can be set so that existing government systems in each band comply with the requirement without any change, thus imposing no cost on existing government users.

Regulators may raise these limits from over time in order to drive more intensive spectrum use. In new assignments, the NTIA should define receiver interference limits in addition to the currently defined transmission permissions (e.g. transmit power or resulting field strength, geographical constraints).

In order not to change the rights of neighboring incumbents, interference limits would be an upper bound on the interference caused by existing operations; thus, transmissions by incumbent neighbors would not exceed the interference limit, and would not trigger a harmful interference claim.

Likewise, transmission permissions would be chosen so that the resulting signals generated by the new operator did not exceed the receiver protection levels of the incumbent neighbors. If the band(s) adjacent to a new assignment were currently quiet, and more intensive use was foreseen, the NTIA could set interference limits that would allow more intensive use in the future, putting the new assignees on notice that they could not depend on the absence of adjacent channel interference to continue into the future.

Receiver interference limits could be added to existing assignments in already-allocated bands, with values chosen to grandfather in existing devices and operations, i.e. to ensure that interfering signals from other operations would not cause harmful interference to incumbent equipment. That is, the interference limits would be low enough, and thus the protection offered would be substantial enough, that interfering signals from other operations would not cause harmful interference to incumbents'current equipment.

New equipment would have to be at least as tolerant of interference as the installed base. These limits would be determined in cooperation with agencies and suppliers, and defined by measurements of actual equipment in use today.

Non-Federal (FCC) Assignments

Within non-Federal licensed bands where the transmitters and receivers are under the control of licensees that use similar technology, have symmetrical interests, and interact with each other repeatedly, economic incentives ensure proper receiver operation. For example cellular systems are all designed to work with the expected interference from other cellular systems within the band.

This allows the carriers to improve their spectrum efficiency within the band by maximizing the number of users, given the interference environment. The primary problem in non-Federal spectrum is at the edges between spectrum allocations. In new allocations, the FCC would follow the same procedure as recommended for new NTIA assignment in the previous section. In cases where devices were initially deployed adjacent to spectrum bands that were quiet, there was no need (technical, commercial or regulatory) to handle interference from an adjacent band; low cost receivers could be built with poor adjacent-band selectivity with no negative impact on device performance.

However, this situation limits the potential value of radio operation when more intensive coexistence becomes necessary. Systems deployed later in adjacent bands have out of band emissions limits defined by their license, and an obligation not to operate in a way that causes harmful interference.[160] But in reality they have an additional de facto requirement placed on them to co-exist with receivers in the adjacent bands which were initially designed to operate in a fairly quiet spectrum neighborhood.

The specifics of this requirement are often not known in advance: the receivers in legacy bands are not required to publish their adjacent band selectivity, nor is there a mechanism for disseminating this information were they interested in providing it.

The designer of the new radio system therefore has little information about the actual out of band emission limit that will be required to prevent interference with pre-existing receivers in the adjacent band. It is therefore difficult to define the technical requirements and determine the cost of building a network next to a band containing pre-existing decoupled receivers. The cross-allocation interference problem is particularly acute where one allocation employs decoupled receivers (i.e. devices not controlled by the spectrum licensee; see Section 4.2 above). Two problems arise: changing usage of adjacent bands and the lack of a well-defined counterparty to coordinate with since the devices are not under the control of the spectrum licensee.

While defining receiver interference limits is a necessary requirement for effectively managing receivers, it may not be sufficient in the decoupled receiver case, particularly where there are millions of consumer-deployed devices. Additional measures may be required to certify that such devices can operate successfully within the receiver interference set for such bands.

As described above, the FCC may elect to add progressively more stringent device performance requirements, from warranties of fitness for purpose and self-certification to the requirement to certify compliance with more or less detailed receiver standards created by a standards body approved by the regulator.

Just as in the Federal case discussed above, receiver interference limits could be added to existing operating rules in already-allocated bands, with values chosen to grandfather in existing devices and operations. In cases where the FCC deems that the protection is more generous than the optimal level, it could put licensees on notice that the interference limits would be reduced in future.

Federal and Non-Federal Assignments Sharing the Same Band

To address the trust issue arising from the risk of adverse possession by non-Federal systems (cf. the "garage door problem" described in Section 3.2), we recommend that the NTIA specifies the interference limits to be tolerated by non-Federal devices sharing with existing incumbent Federal users within each band, and that the FCC specify the appropriate device requirements that will ensure adequate operation of non-Federal devices. This case falls into the device certification column in the Federal/non-Federal sharing in Table D.2.

The requirements will differ from band to band based on the characteristics of the Federal and non-Federal systems.

The FCC should create an equipment authorization process for future non-Federal devices designed to share Federal spectrum to confirm that they can operate within the receiver interference limits for the band of interest. If the non-Federal devices are unlicensed, FCC-specified receiver performance standards may be required; if the non-Federal systems are licensed, with receivers under the control of a licensee, self-certification of adequate performance may be sufficient.

The receiver interference limits, and associated authorization process, assure that non-Federal devices are able to withstand the interference from Federal systems before they are deployed.

Improving the Performance of Receivers

Setting the level of interference that receivers have to deal with represents a choice in the balance of rights between transmitters and receivers in adjacent bands. More allowed interference (i.e. higher receiver interference limits) favor adjacent transmitters since higher power allows increased performance through increased range and higher data throughput, and disfavors receivers since they have to tolerate lower signal-to-noise ratios or invest in better filters to reject this noise. Conversely, less interference (lower limits) tips the balance in favor of receivers over transmitters.

For practical reasons, it is likely that protection limits will be set in the first instance to affirm the status quo. Thus, the balance between transmission and reception rights will probably not be optimal, and will need to change to maximize social welfare. In cases where bargaining across a band boundary is feasible, e.g. when there are few parties, well-defined rights and transaction costs are low, one can expect that this adjustment will happen through negotiation.

More often than not, this will benefit incumbents, since reaching the optimum will entail that the receiver interference limit (i.e. allowed out-of-band interference) will need to increase; and thus, the transmitter will pay the receiver. However, everyone will be better off since the transmitter would only willingly pay the receiver less than they would gain overall from the new regime.

In some cases, e.g. where parties on either side of the boundary have symmetrical interests (everybody's both a receiver and transmitter, with similar cost/benefit structures), the adjustment in protection limits can occur without a wealth transfer.

Unfortunately the costs of negotiation are often prohibitively high in wireless, due to large numbers of rights-holders, lack of information, poorly defined rights, and strategic behavior. In such cases, the government has to make the economic efficiency calculus and adjust receiver interference limits (and thus operator's rights) unilaterally. The shift in the cost burden from transmitters to receivers can be staged over time. The regulator would give advance notice that interference limits will be increased, e.g. indicating that it will increase might increase from its initial value of X dB(pV/m)/MHz to X + 5dB in ten years, and X + 15 dB in fifteen years. As increasing interference limits allow more energy to be transmitted in adjacent bands, receivers will have to maintain their performance levels.

In cases where the NTIA or FCC has decided to go beyond setting interference limits, e.g. by prescribing minimum front-end filter requirements for receivers, it may tighten these requirements from time to time.

However, obtaining the maximum value from radio operations is a system optimization challenge, both in engineering and economics. It will not always be the case that the optimum solution requires improving receiver quality; it may be most efficient to deploy poorer receivers if an operator obtains bands to choose from through spectrum sharing, and can cope with interference degradation by hopping to another channel.

APPENDIX E: RESOLUTION OF SPECTRUM SHARING DISPUTES

As spectrum sharing grows, disputes among both active and potential users in the Federal and non-Federal sectors are likely to become more common. Section 5.2 noted the importance of having a robust framework in place to resolve such disputes. Indeed, the lack of such a framework will likely prove an impediment to a growth in sharing. This Appendix sets out in more detail the necessary features of such a framework, and offers some suggestions for how it might be put into practice.

Disputes should be viewed as normal because we expect them to arise from time to time in the course of spectrum sharing activity, but also infrequent, since the purpose of an effective system is allow sharing to occur routinely. For those disputes that do arise, a suitable resolution process for spectrum sharing rights would have a number of desirable features:

- **Dependable:** While the outcome of any particular dispute will depend on case-specific facts, the process and jurisdictional responsibilities for resolving disputes should be clear to all parties. Ambiguity as to who has ultimate authority to resolve disputes or about the options for escalating the resolution of disputes, can adversely impact behavior before and after disputes arise and before any final resolution process is actually implemented.
- **Timely:** It is also important that disputes be resolvable relatively quickly and in a predictable time frame. The possibility of disputes dragging on without resolution for an indeterminate amount of time imposes significant business risk on any sharing model. Real operational costs are incurred while the dispute remains unresolved, and uncertainty as to the ultimate outcome is a deadweight drag on prospects for future returns. Together the threat of such costs reduce ex ante incentives to engage in sharing, and increase ex ante incentives to invest in deadweight bargaining costs.
- **Efficient:** disputes should be resolvable with as little additional cost as possible, and ideally ought to maximize total surplus. From the perspective of designing an efficient dispute resolution process, it is more critical to obtain an acceptable resolution at low cost and with a predictable (preferably short) timing than it is to obtain perfect resolution on a case-by-case basis.

Today, we are a long way from having a suitable framework for resolving spectrum disputes and enforcing decisions. Jurisdictions are mixed and ambiguous: the FCC is responsible for managing non-Federal, including commercial licensees, and the NTIA is responsible for managing Federal authorizations. Since a very significant share of spectrum is already shared between Federal and civil users, jurisdictional disagreements occasionally arise between FCC and NTIA. An additional complication is that the NTIA is part of the Executive branch, whereas the FCC is an independent regulatory authority whose ultimate mandate and jurisdictional authority derives from Congress and the Communications Act of 1934. In light of the long term goal of blurring the boundary between non-Federal and Federal users from the perspective of interference management, it may be desirable to aspire to a future in which there is a single and focused jurisdictional authority with the sole responsibility of adjudicating and enforcing spectrum usage rights. In this ideal world, there would not be separate agencies administering Federal and civil spectrum licenses. The resolution of interference claims would not depend on who held the license but on the merits of the interference claim, which would be principally technical. The rationale for this perspective is that the ultimate goal is a world in which spectrum may be efficiently and dynamically reallocated while users are protected against harmful interference. A receiver experiences interference in a given RF environment regardless of who owns the license, but, at present, the right to prosecute such a claim may depend on the assignment of access rights. If we want to create a world in which such access rights might be dynamically altered (transferred, shared, or redefined) though market-based sharing, then we need a framework that is independent of the actual assignment. Finally, because technology evolves over time (e.g., the measurement capabilities of spectrum sensing technologies or tunability of transmitter or receiver systems affects the granularity of feasible rights assignments), the interpretation and management of interference disputes will not be purely technical but will likely also require some legal (interpretation of current license terms) and economic (cost implications of alternative dispute outcomes) expertise as well. However, the scope for extending beyond a narrow focus on offering predictable resolution of narrow sharing-related interference disputes needs to be tightly circumscribed.

Detailed specification of such a framework goes beyond the scope of this report. PCAST nonetheless recommends that the importance and need for a dependable, timely, and efficient institutional framework for resolving future spectrum disputes be formally acknowledged. Such a framework would have many parts (clear definitions of property rights, new infrastructure such as the spectrum data bases to track and support enforcement of spectrum usage rights, and institutional reforms to support better and more predictable enforcement of spectrum usage rights, including sharing rights). Options may include a variety of future alternate dispute resolution procedures, including creation of a "GSA-like" spectrum court. In the interim, the White House Spectrum Management Team (SMT) can serve as the Federal adjudicator of spectrum sharing disputes.

This role might be permanently established, or replaced by other structures as the needs for the system are better understood through experience. The SMT's role would embrace several points:

- The SMT will be final arbiter for disputes of spectrum usage rights when Federal users have primary access rights;
- The SMT that it will cede authority to (will defer to) the FCC in all spectrum where either the Federal users are secondary or co-primary with the FCC;
- The SMT will commit on behalf of the NTIA to provide better transparency and access to data to allow all parties better access to information relevant to preventing and resolving spectrum sharing disputes;
- The SMT/NTIA will promulgate publicly its policies and framework for managing spectrum usage rights and adjudicating outcomes;
- The SMT should establish a sharing review process to make recommendations to the NTIA and FCC when jurisdictional disputes arise that impede progress toward sharing spectrum.

APPENDIX F: ILLUSTRATIVE LIST OF INTERNATIONAL RESEARCH PROGRAMS IN ADVANCED SPECTRUM USE

Countries	Project	Description/Goal	Reference Link
China	Program 863	Broad S&T improvement project, with a $200 billion allocation. Focuses in part on spectrum sensing and allocation; demonstration of dynamic spectrum sharing in 694-806 MHz band.	www.most.gov.cn/eng/programmes1/200610/t20061009_36225.htm
China	Program 973	National Basic Research Program; Researches key techniques for efficient spectrum utilization.	www.973.gov.cn/English/Index.aspx
EU	COST (Cooperation in Science and Technology) IC0902: Cognitive Radio and Networking for Cooperative Coexistence of Heterogeneous Wireless Networks	Technical multi-country collaboration on Cognitive Radio (CR) impact on all layers of the protocol stack (algorithms and protocols).	www.cttc.es/en/project/091029-cost-ic0902.jsp
EU	COST–TERRA (ICO 905): Techno-Economic Regulatory Framework for Radio Spectrum Access for Cognitive Radio/Software Defined Radio (TERRA)	Deployment scenarios, business analy- sis, regulatory impact COST Actions established in the area. IC0902 looks into the technical aspects of cognitive radios and networks. TERRA looks into the policy and economic aspects.	www.cost.eu/domains_actions/ict/Actions/IC0905
EU	E2R / E3: End to End Efficiency	CR system for heterogeneous networks; integration into cellular / cognitive pilot channel.	ict-e3.eu/
EU	COGEU: Cognitive EU	Secondary spectrum trading and the creation of new spectrum commons regime.	www.ict-cogeu.eu/
EU	CREW: Cognitive Radio Experimental World	Federation of testbeds of CR with heterogeneous systems.	www.crew-project.eu/
EU	QoSMOS: Quality of Service / Mobility Driven Service	Managed QoS (Quality of Service) in mobile broadband in mixed licensed spectrum.	www.ictqosmos.eu/
EU	FARAMIR: Flexible and spectrumAware Radio Access through Measurements and modeling In cognitive Radio systems	The goal of the FARAMIR project is to research and develop techniques for increasing the radio environmental and spectral awareness of future wireless systems.	www.ictfaramir.eu/
		The project aims to accomplish this by developing a reference architecture and implementation for Radio Environment Maps (REMs), which are essentially knowledge bases in which cognitive radios and store and access information on the environment and other wireless systems. New spectrum sensing technologies and algorithms are also being developed, including novel radio neighborhood mapping techniques for characterizing interference sources.	
EU	ARAGORN: Adaptive Reconfigurable Access and Generic Interfaces for Optimization in Radio Networks	Collaborative intelligence for Industrial, Science, and Medical (ISM) band.	ictaragorn.eu/

Countries	Project	Description/Goal	Reference Link
EU	SAMURAI: Spectrum Allocation and Multi User MIMO: RealWorld Impact	An industrially focused consortium composed of a telecommunication network equipment maker, chipset vendors, test equipment vendor and universities. The focus of the project is to tackle the challenge of next generation telecommunication systems using multi user MIMO (Multiple Input / Multiple Output) and aggregated spectrum techniques.	www.ictsamurai.eu/
EU	ACROPOLIS: Advanced coexistence technologies for radio optimization in licensed and unlicensed spectrum	Advanced coexistence technologies for radio optimization in licensed and unlicensed spectrum.	www.ictacropolis.eu/
U.S. (DOD)	NRL Cognitive Radio Test Laboratory	Utilizes Virtual Field Test (VFT) environment, capable of survey and capture of reallife spectrum with bandwidth up to 40 MHz and RF up to 4.6 GHz. The stored digitized data can be manipulated and injected into the testbed to emulate realistic electromagnetic scenarios.	www.nrl.navy.mil
		Beyond simulation, the testbed could serve as a baseline of canonical scenarios to evaluate performance of competitive CR/DSA technologies.	
U.S. (DOD)	ORBIT: Open Access Next Generation Wireless Network Testbed	Twotier laboratory emulator/field trial network testbed designed to achieve reproducibility of experimentation, while also supporting evaluation of protocols and applications in realworld settings. The core of the wireless network emulator is based on a large twodimensional grid that enables remote users to conduct reproducible networking experiments with large numbers of programmable wireless nodes.	www.orbitlab.org
U.S. (DOD)	Public Safety Communications Research Lab (PSCR)	PSCR conducts a broadbased technical program to facilitate communications interoperability and information sharing among wireless and IT systems within the public safety and homeland security community. Technical thrusts within the program include: Project 25 Standards Development, Project 25 Compliance, Public Safety Audio Quality, Public Safety Video Quality, Public Safety Interoperability Test Tools, Public Safety Broadband Communications.	www.pscr.gov/
U.S. (DOD)	Cognitive Radio Network Testbed (CORNET)	A collection of high performance servers, unique and extremely flexible RF hardware, and softwaredefined radio architectures. CORNET has 48 SoftwareDefined Radio nodes, RF portion based upon the Universal Software Radio Peripheral 2, custom developed USRP2 daughterboard based on the Motorola RFIC4, high performance servers for signal processing tasks, and distributed node topology throughout the Virginia Tech ICTAS research building.	cornet.wireless.vt.edu
U.S. (NSF)	CIF: NeTs Small: Collaborative Research: Distributed Spectrum Leasing via CrossLayer Cooperation	CIF: NeTs Small: Collaborative Research: Distributed Spectrum Leasing via CrossLayer Cooperation	nsf.gov/awardsearch/showAward.do?AwardNumber=0914912

(Continued)

Countries	Project	Description/Goal	Reference Link
U.S. (NSF)	CIF: NeTs Small: Collaborative Research: Distributed Spectrum Leasing via CrossLayer Cooperation	This research project focuses on two significant technical and regulatory issues which must be resolved to ensure successful deployment of cognitive radio networks. The first issue relates to the network's ability to manage interference in a distributed fashion without cooperation from the Federal Primary Access users. Here, the research tasks include the analysis, from a signal processing and algorithmic point of view, of various price-based schemes for dynamic spectrum allocation in a broad range of CRN scenarios under a variety of regulatory restrictions. The second relevant issue pertains to the design of a secondary market for the spectrum. The research investigates the analysis of various design choices taking into account specific spectrum sharing techniques and the associated behavior of sellers (i.e. Federal Primary Access users/primary network service providers (NSP)) and buyers (i.e. cognitive users).	nsf.gov/awardsearch/showAward.do ?AwardNumber=1017982
U.S. (NSF)	NeTS: Small: A Practical and Efficient Trading Platform for Dynamic Spectrum Distribution	This project develops STRADE, an auctiondriven spectrum trading platform to implement the spectrum marketplace. STRADE differs significantly from conventional FCCstyle spectrum auctions that target only a few large corporate players and take months or years to conclude. Instead, STRADE serves many small players and enables on-the-fly spectrum transactions. In essence, STRADE selectively buys idle spectrum pieces from providers and sells them to a large number of buyers matching their individual demands. By effectively multiplexing spectrum supply and demand in time and space, the proposed marketplace also significantly improve spectrum utilization.	nsf.gov/awardsearch/showAward.do ?AwardNumber=0915699
		The design of STRADE focuses on achieving spectrum multiplexing/reuse to improve spectrum utilization while guaranteeing economic robustness to encourage player participation and minimize market manipulation. This project focuses on tightly integrating novel algorithms of dynamic spectrum allocation with economic mechanism design.	

Appendix G: The Test City and Mobile Test Service

G.1. Introduction

Chapter VI called for the creation of an urban Test City complemented by a Mobile Test Service to support rapid experimentation and gain essential operational test data to establish the dependability of both the technology and the management techniques supporting the new spectrum architecture. This Appendix provides further technical detail on these test facilities and expands on the reasons creating them.

The Test City and Mobile Test Service will be needed by various groups: DOD and other Federal wireless users, public safety users, and numerous commercial entities including both wireless equipment manufacturers and service providers.

These entities individually cannot afford their own dedicated test equipment, support staff, and other logistical resources to implement large scale testing, so it is logical to share the unique urban Test City facility and associated costs for the infrastructure, instrumentation, tools, software, and technical and operational support personnel.

Moreover, the facilities will evolve continually by incorporating newly developed systems and even early research concepts.

Finally, the availability of large scale dynamic spectrum testing facilities will provide broad visibility to the benefits of emerging spectrum sharing technology, demonstrating improvements in cost, performance, reliability, spectrum efficiency, innovative applications, and network management practices applicable to a broad array of organizations in the business, technical, and regulatory worlds that would otherwise be unaware of such advancements.

G.2. What Testbeds Currently Exist? What Impact Have They Had?

In 2004, President Bush directed NTIA to implement the recommendations in two Department of Commerce reports on spectrum that included the recommendation for a Spectrum Sharing Innovation Test-Bed.

The stated goal of that testbed was "to objectively evaluate new technologies to facilitate sharing between federal and non-federal spectrum users. If sharing is successfully demonstrated, the results of the Test-Bed can be used as the basis to establish service rules for the technologies that have operated in the Test-Bed frequency bands."[161]

The Cognitive Radio Test Laboratory and Tactical Edge Network Test Bed, both under the direction of the Naval Research Laboratory in Washington, D.C., along with the Spectrum Sharing Innovation Test-bed and the Public Safety Communication Research Lab (PSCR) in Colorado, both under the direction of NIST through the Institute for Telecommunications Sciences (ITS), provide this function.

By complementing industry advances, these facilities have led to various technological developments that now make it possible to test the implementation of sharing in larger simulated real-world environments.

Other government facilities capable of handling dynamic spectrum sharing testing exist at Idaho National Laboratory (INL), Oak Ridge National Laboratory (ORNL), and at various DoD test ranges.

There are a number of existing university based facilities as well, such as the ORBIT facility maintained by the Wireless Information Network Lab (WINLAB) at Rutgers University, in which several nodes in a room are interconnected with each other running controlled experiments. The Virginia Tech Cognitive Radio Testbed Network facility (CORNET) has software defined wireless nodes that are spread throughout a building, along with a few nodes outside the building that are also interconnected with each other for controlled experiments, yet operate in a real-world setting.

CentMesh at North Carolina State University is composed of open source Wi-Fi device covering the campus, offering the potential for experimentation for improving Wi-Fi to be more efficient and customized to a region. This is an especially important contribution as future networks will be heterogeneous relying on Wi-Fi combined with commercially operated network.[162]

A number of agile radio test beds are available, and while they can and will be useful for spectrum sharing R&D, there is no current test bed capability with the availability, capability, size, and support infrastructure to provide the comprehensive testing resource needed to develop a new national spectrum management regime and insure both business interests and legacy users' needs are met.

G.3. Additional Features of the Mobile Test Service and the Test City

What are Some Examples of Systems that Might Be Tested?

The extension of today's emerging White Spaces based wireless systems into the broader and more complex dynamic shared spectral domains offers immediate and compelling uses for the Test City. Applications include video surveillance systems (the first certified application), Super Wi-Fi systems, or novel systems with associated applications that simply do not exist today. One of the first applications for the use of the Mobile Test Service might be to ferret out the real opportunities for dynamic sharing in the 1755–1850 and 3550–3650 MHz spectral ranges by enabling carefully constructed experiments with dynamic sharing system prototypes designed to operate in these frequency bands under a variety of different geographical and signal environments. This could provide much enhanced confidence in the ability for the systems to peacefully co-exist and/or the opportunity to identify system design flaws that would need to be corrected to enable successful shared spectrum operation. This capability can then be applied in subsequent experiments to demonstrate such applications as monitoring for rural transportation systems, public safety spectrum sharing in rural locations, and massive sensor systems for environmental and agricultural monitoring.

What Should Be Tested in a Test City Besides Systems?

In some ways testing of physical hardware or even systems is one of the least important aspects of the dynamic spectrum sharing testbed. The long lasting value is in the knowledge, skills and relationships that are developed through building and operating the testbed. Important factors include:

1. Knowing how to work with NTIA, DOD, FCC, Defense Information Systems Agency, Department of Justice, etc. to insure that testing can be done without disrupting essential operations.

2. Developing well defined test patterns that can be used to carry out direct comparisons with various systems. Experience gained by using the Mobile Test facility to study a particular location with a unique set of spectrum occupancy and usage patterns will be invaluable for understanding real-world issues that dynamic spectrum sharing systems will encounter.

3. Identifying and developing methods to train current and future personnel in various areas:

a) Personnel for large scale test, such as students or military;
b) Personnel who can maintain and install equipment as needed;
c) Software support for tool maintenance and data collection;
d) Experiment coordinator and staff who by proper planning and execution of experiments will play a critical role in providing reassurance to incumbent stakeholders;
e) Application test personnel who are sensitive to the importance of the human aspects of a system (latency, data rate, error rate, etc.) that contribute to a positive, acceptable, or unacceptable user experience under various application environments and related system load conditions;
f) Operations staff (legal, safety, administrative).

4. Developing various software resources:

a) Emulation software to perfect experiments before getting to testbed;
b) Automated testing tools;
c) Deployment tools;
d) Database development and management skills to support the capture and structured dissemination of the test results;
e) Software for measuring, analyzing and reporting experimental results.

How Can Existing Resources Be Leveraged?

Numerous resources exist today that can complement the Test City and Mobile Test Service. Ideally a network of test capabilities will emerge over time to connect and positively extend the disparate capabilities of individual test environments. As a specific example, a lab facility is needed to provide controlled equipment experiments to spot and minimize the problems that might later be encountered when the system enters the Test City test regime.

The testing resources can be leveraged through remote connections to the database and computational facilities of the existing network testbeds described above. In particular, the NIST Public Safety Communications Research program in Boulder, Colo., provides a common field environment for manufacturers, carriers, and public safety agencies to test and evaluate advanced broadband communications equipment and software tailored specifically to the needs of emergency first responders. Facilities such as those at the INL can provide commercial grade cellular base stations for experimentation with variations of commercial cellular standards. NRL Cognitive Radio Test Laboratory and NRL Tactical Edge Network

Test Bed offer the potential for testing against classified signals and systems and thus can augment publically available facilities. INL provides a vast expanse for testing that can uniquely support certain types of testing, for example INL supports testing against live aircraft wireless systems today. Army C4ISR and Radio Analysis and Experimentation Facilities are well suited to experiment with policy issues and military systems.

In addition to a lab facility, a simulation environment is needed to develop models for sharing systems. The radar simulation lab at ORNL, with its ability to simulate a number of military radar systems, is perhaps the best example. This simulation capability should be greatly expanded both in computational capacity and range of situational models so that it can incorporate results from early experiments and measurements using the dynamic spectrum testbeds. A comprehensive environment will need to adequately represent the intricate inter-dependencies between the wireless communication networks that carry data traffic, the social network that generates the load, and the market. In other words, the social and urban context needs to be adequately represented to understand the pros and cons of various spectrum-sharing technologies, both in the short term and the long term. Such a modeling environment will assist policy analysts, commercial vendors and individual researchers interested in dynamic spectrum access and trading; location-aided services; impact of disruptive technological changes on the future use of spectrum (e.g. the advent of next generation smart phones and tablets); and communications system design and analysis for large-scale natural or human-initiated crises.

Furthermore, the expanded simulation lab can also supplement experiments in the dynamic spectrum testbeds by providing simulated environments to augment real-world situations in order to further explore the capabilities of hardware and algorithms. The computational modeling environment could be coupled to the mobile test facility to enhance these models, calibrate them, and validate their conclusions. Researchers have taken initial steps to build such a modeling environment.[163]

G.4. How Much Will the Test City and Mobile Test Service Cost?

Test City: Building a Test City is an even more ambitious undertaking than building an urban wireless network. Not only does the Test City need to provide the coverage that an urban network does, but it has to provide the flexibility to easily and cost effectively test new standards in new spectrum on new small cell architectures while supporting testing with large user bases and multiple sources of potential interference. Building a Test City from scratch would be prohibitively expensive. Leveraging resources and funding in a public-private partnership brings the cost down significantly and insures participation of both government and commercial entities that is essential to the success of spectrum sharing. The most valuable contribution that a state or local Test City government can make to the Test City is the contribution of resources, such as fiber, small cell site locations (e.g. utility poles and rooftops), and facilities to house equipment and people. A good list of potential public resource contributions to a Test City can be found in the Seattle RFI for public safety.[164] These contributions offset the major costs in building a network (the site lease and backhaul costs) and support a wide area small cell deployment. In addition to these physical assets, government assistance in the zoning and siting process for small cells would both reduce costs and speed time to deployment.

Assuming a deployment architecture along the lines of a Distributed Antenna System (DAS) architecture, major remaining CAPEX (capital expenditure) components of the test city system are the processing infrastructure for both the network signal processing as well as the control and application framework. There is also significant pre-launch OPEX (operational expenditure) as sites are brought online that must be taken into consideration in the CAPEX budget. The contribution of fiber also offsets the backhaul costs, a major OPEX component of running any network. The major remaining OPEX components of the Test City are personnel, utilities, and maintenance. By far the largest of these is the personnel costs, owing to the broad scope of activity within the Test City. A core of engineers and technicians is needed to operate and maintain the network. In addition there needs to be a small team to coordinate spectrum activities not only with FCC and NTIA but also with local incumbent entities participating in the sharing experiments. The Test City team needs an experiment coordinator with a team that includes staff researchers to assist entities bringing in systems to test in the Test City. This team is essential for integrating and disseminating knowledge regarding dynamic spectrum testing procedures, policies and lessons learned. Finally the Test City team should have a software engineering team responsible for maintain and extending the test sharing databases to incorporate new metrics and algorithms into sharing experiments. This database should be maintained as an open source resource to serve as a reference for implementation and extension of the spectrum access system. An example three year Test City budget based on the above assumptions is provided in Table G.1.

Table G.1. Estimated Capital and Operational Expenditures for the First Three Years of a Test City

Financial Summary w/ Public Private Partnership	Year 1	Year 2	Year 3
End of Year Sites	$200	$250	$250
OpEx			
G&A	$5,130,000	$5,066,000	$5,072,000
Site OPEX	$448,000	$2,400,000	$2,400,000
Core Network	$783,000	$420,000	$420,000
OpEx Total	$6,361,000	$7,886,000	$7,892,000
CAPEX	$11,975,000	$2,860,000	–

Mobile Test Service: We propose a total of 12 "cells on wheels" (COWs) for the Mobile Test Service. A single commercial COW costs roughly $180,000, but the inclusion of additional test and measurement equipment would likely bring that cost to about $300,000 per COW, for a vehicle CAPEX of $3.6 million. Annual replacement cost is budgeted at 10% of initial capital costs. Operational costs for the 12 COWS include expenses for drivers, and onsite expenses for backhaul, utilities, and general maintenance and support. The intent is for this to be a lightweight capability that can be used by any of the existing testbeds to facilitate testing in remote locations, with a small staff for maintenance and operation of the Mobile Test Service. To save costs, the personnel associated with moving the Service are supplemented with contractors for the move itself. The CAPEX and OPEX required to support creation and three years of operation of the Mobile Test Service are summarized in Table G.2, for a total three year cost of $11.6M.

Table G.2. Estimated Capital and Operational Expenditures for the First Three Years of the Mobile Test Service

Mobile Test Service	Year 1	Year 2	Year 3
OpEx			
G&A	$1,085,000	$1,150,100	$1,219,106
Operation Costs	$998,000	$1,057,880	$1,121,353
Network	$80,000	$84,800	$89,888
OpEx Total	$2,163,000	$2,292,780	$2,430,347
CAPEX	$3,960,000	$396,000	$396,000

LIST OF ABBREVIATIONS

2G, 3G, 4G	second, third, fourth generation
ADC	Analog to Digital
ADSL	Asymmetric Digital Subscriber Line
AMP	Advanced Manufacturing Partnership
ASA	Authorized Shared Access
AWS	Advanced Wireless Services
EIRP	Effective Isotropic Radiated Power
CAPEX	Capital Expenditure
CEPT	European Conference of Postal and Telecommunications Administrations
CORNET	Cognitive Radio Testbed Network (Virginia Tech)
COW	Cell on Wheels
CSEA	Commercial Spectrum Enhancement Act
CSMAC	Commerce Spectrum Management Advisory Committee
CTO	Chief Technology Officer
DARPA	Defense Advanced Projects Research Agency
DAS	Distributed Antenna System
DFS	Dynamic Frequency Selection
DNS	Domain Naming System
DRM	Digital Rights Management
DSA	Dynamic Spectrum Access
DOD	Department of Defense
EU	European Union
FAS	Frequency Assignment Subcommittee (IRAC)
FCC	Federal Communications Commission
FDD	Frequency Division Duplexing
FSMS	Federal Spectrum Management System
GAO	Government Accountability Office
GPS	Global Positioning System
GSA	General Services Administration
GSM	Global System for Mobile Communications
HETNET	Heterogeneous Network

INL	Idaho National Laboratory
ITS	Institute for Telecommunications Sciences (NTIA)
ITU	International Telecommunication Union
IRAC	Interdepartment Radio Advisory Committee
LSA	Licensed Shared Access
LTE	Long Term Evolution
M2M	Machine to Machine
MIMO	Multiple-input multiple-output
MNO	Mobile Network Operator
NEC	National Economic Council
NIST	National Institute of Standards and Technology
NRC	National Research Council
NRL	Naval Research Laboratory
NSS	National Security Staff
NSTC	National Science and Technology Council
NTIA	National Telecommunications and Information Administration
OECD	Organisation for Economic Cooperation and Development
Ofcom	Office of Communications (UK)
OMB	Office of Management and Budget
OPEX	Operational Expediture
ORNL	Oak Ridge National Laboratory
OSTP	Office of Science and Technology Policy
PCAST	Presidents'Council of Advisers on Science and Technology
PCS	Personal Communications Services
PPP	Public Private Partnership
PPSG	Policy and Plans Steering Group
PSBN	Public Switched Broadband Network
PSCR	Public Safety Communication Research Lab
PSTF	Public Safety Trust Fund
RF	Radiofrequency
RLEC	Rural Local Exchange Carriers
SAS	(Federal) Spectrum Access System
SMR	Specialized Mobile Radio
SMT	Spectrum Management Team
SPTF	Spectrum Policy Task Force
SRF	Spectrum Relocation Fund
STA	Special Temporary Authority
SSP	Spectrum Sharing Partnership Steering Committee
TAC	(FCC) Technology Advisory Council
TDD	Time Division Duplexing
TTL	Time to Live
WS	White Spaces database technology, initially created for TV Bands
UNII	Unlicensed National Information Infrastructure
UTC	Utilities Telecom Council
WAPECS	Wireless Access Policy for Electronic Communications Services
WIN	Wireless Innovation Fund

WINLAB	Wireless Information Network Lab (Rutgers)
WISP	Wireless Internet Service Provider
WISPA	Wireless Internet Service Providers Association
WRC	World Radiocommunication Conference
XG	NeXt Generation

GLOSSARY

Adaptive radios: General name for devices that can modify their operation in various ways to obtain optimum performance in the presence of changing conditions and unpredictable interference from other systems.

Advanced Wireless Service (AWS): The collective term used for new and innovative fixed and mobile terrestrial wireless applications using bandwidth that is sufficient for the provision of a variety of applications, including those using voice and data (such as Internet browsing, message services, and full-motion video) content. Advanced wireless systems could provide, for example, a wide range of voice, data, and broadband services over a variety of mobile and fixed networks. www.federalregister.gov/articles/ 2006/05/24/06-4769/advanced-wireless-services#p-7.

Allocation: Frequency bands of the electromagnetic spectrum are allocated for the purposes of use by one or more (terrestrial or space) radiocommunication services or the radio astronomy service under specified conditions. The FCC's Table of Frequency Allocations consists of the International Table of Frequency Allocations and the United States Table of Frequency Allocations. The FCC's Table of Frequency Allocations is codified at Section 2.106 of the Commission's Rules. These resources designate the particular areas of the electromagnetic spectrum. www.ntia.doc.gov/files/ ntia/publications/6_5_11.pdf, www.fcc.gov/topic/frequency-allocation.

Architecture: The set of technical rules and regulatory practices governing the operation of wireless systems across the entire radio spectrum.

Assignment: Authorization for a radio station to use a radio frequency or radio frequency channel under specified conditions. www.ntia.doc.gov/files/ ntia/publications/6_5_11.pdf.

Auctions: The FCC uses auctions (competitive bidding) as one of the primary means of choosing among two or more mutually exclusive applications for an initial license for most commercial services, including wireless, television, and radio. In a spectrum auction, parties apply to become qualified bidders for one or more spectrum licenses and take part in an online auction for those licenses. By using auctions, the FCC seeks to award licenses to those who value them most and who will have an incentive to use them most effectively. Prior to Congress granting the FCC auction authority in 1993, the Commission relied upon comparative hearings and lotteries to select a licensee from mutually exclusive applicants. www.fcc.gov/topic/auctions.

Backhaul: General term for the connections from the core of a network to the outer elements that deliver service. In a cellular phone network, for example, the link from an individual cell tower to the telephone company's core system is part of the backhaul.

"Beachfront" frequencies or spectrum: Highly valued spectrum is sometimes called "beachfront" spectrum. For many mobile radio systems, the 300 MHz to 3 GHz spectrum range is the portion of the spectrum where scarcity concerns are the greatest. However, for some industry representatives, the "beachfront" spectrum is larger, located anywhere between 100 MHz to 6 GHz. As spectrum-dependent technologies improve over time, the definition of high-value spectrum can change. www.gao.gov/new.items/d11352.pdf.

Bluetooth: a technology for enabling secure wireless communications among multiple devices in small areas.

Broadband: high-speed Internet access. www.gao.gov/new.items/ d11352.pdf.

Code division multiple access (CDMA): A system enabling multiple users to transmit and receive at the same frequency by encoding their signals so that they can be detected independently of each other.

Cognitive radios: A cognitive radio is a radio that can change its transmitter parameters based on interaction with the environment in which it operates. transition.fcc.gov/pshs/ techtopics/techtopic8.html.

Distributed Antenna System: A distributed antenna system is a network of spatially separated antenna sites called "nodes" connected to a common source that provides wireless service within a geographic area or structures. fjallfoss.fcc.gov/edocs_ public/attachmatch/ FCC-10-177A1.doc.

Desensitization: (also known as receiver overload): Loss of receiver performance caused by reception of signals on frequencies adjacent to a receiver's nominal operating frequency, leading to detection of spurious signals or detuning of the receiver.

Dynamic Frequency Selection (DFS): System by which transmitters and receivers can autonomously switch to different frequencies within a band to improve transmission and minimize interference to other users. The NTIA and DoD worked with industry to develop DFS technology that allows 5.8 GHz Wi-Fi devices to detect and avoid military radars.

Dynamic Spectrum Aaccess (DSA): System that can actively search for unused spectrum and organize a network of devices and transmitters to operate in it, thereby ensuring that no interference is caused to other users.

Fast Track: NTIA's "Fast Track" report officially titled "An Assessment of the Near-Term Viability of Accommodating Wireless Broadband Systems in the 1675-1710 MHz, 1755-1780 MHz, 3500-3650 MHz, and 4200-4220 MHz, 4380-4400 MHz Bands" evaluated four bands that appeared to lend themselves to rapid decision-making and the possibility that wireless broadband systems could be accommodated within five years. www.ntia.doc.gov/files/ntia/publications/fasttrackevaluation_11152010.pdf.

Femtocell: A femtocell is a low-power access point, based on mobile cellular technology, providing wireless voice and broadband services to customers with a limited range within a home or in an office environment. Femtocells connect to the mobile operator's network facilities via a standard consumer broadband connection, such as DSL, cable or fiber. Data to and from the femtocell is carried over the Internet - or at least, over an Internet technology-based network provided by an Internet Service Provider. The wireless subscriber connects to the femtocell via the normal cellular service technologies just as if he/she were using a conventional macro-cellular network connection. transition.fcc.gov/pshs/techtopics/ techtopics23.html.

FirstNet: The First Responder Network Authority *(FirstNet)* is an independent authority to be formed within NTIA under the 2012 Payroll Tax Agreement to oversee the establishment of a nationwide public safety broadband network. Once established, *FirstNet* will be responsible for taking all actions necessary to ensure the building, deployment and operation of the nationwide public safety broadband network.

Geolocation Database: A Geolocation database includes the physical location of registered services as well as other transmission characteristics.

GSM: Global System for Mobile Communications, originally Groupe Spécial Mobile, is a standard for wireless communications developed to unify the incompatible systems in use in a number of different European countries. Originally a part of 2G (2nd generation) networks, it is now evolving to include LTE Advanced protocols.

Heterogeneous network: A heterogeneous network is a network connecting computers and other devices with different operating systems and/or protocols. For example, local area networks (LANs) that connect Microsoft Windows and Linux based personal computers with Apple Macintosh computers are heterogeneous. The word heterogeneous network is also used in wireless networks using different access technologies. For example, a wireless network which provides a service through a wireless LAN and is able to maintain the service when switching to a cellular network is called a wireless heterogeneous network. en.wikipedia.org/wiki/Hetero geneous_network.

Hidden node: In a wireless network nodes can be hidden from each other if they are in communication with a central station but not directly with each other. The presence of hidden nodes creates complications for transmitting throughout the network, and also when two or more networks are trying to share the same frequency and physical space.

Interference: Interference is any unwanted radio frequency signal that prevents you from watching television, listening to your radio or stereo or talking on your cordless telephone. Interference may prevent reception altogether, cause only a temporary loss of a signal, or affect the quality of the sound or picture produced by your equipment.www.fcc.gov/topic/inter ference.

License: A "license" is a document issued by the relevant authority authorizing the use of a radio station or equipment and/or radio frequencies to provide electronic communication services under standard conditions (a class license) or authorizing the construction ownership and exploitation of an electronic communication network or service when the number of such networks or services must be limited and specific conditions of use are attached (individual rights of use). ec.europa.eu/information_society/policy /ecomm/radio_spectrum/_document_storage/studies/shared_use_2012/scf_study_shared_spectrum_access_20120210.pdf.

Licensing: The FCC is responsible for managing and licensing the electromagnetic spectrum for commercial users and for non-commercial users including: state, county and local governments. This includes public safety, commercial and non-commercial fixed and mobile wireless services, broadcast television and radio, satellite and other services. In licensing the spectrum, the Commission promotes efficient and reliable access to the spectrum for a variety of innovative uses as well as promotes public safety and emergency response. www.fcc.gov/topic/licensing.

LTE (Long Term Evolution): A standard for wireless communication that upgrades 3G (3rd Generation) systems to increase transmission speed and capacity. LTE Advanced is compatible with 4G (4th generation) standards.

Machine-to-machine (M2M): Machine-to-machine communications and devices refer to systems of semi-autonomous components that exchange data on wired or wireless networks. Examples include smart-grid meters that report electricity usage or medical monitoring devices that report critical health data.

Multiple-input multiple-output (MIMO): A form of smart antenna technology that uses multiple antennas for both the transmitting and receiving ends of a wireless communication system to improve data rate and reliability, especially in complex urban areas where such systems can make use of multiple signal paths to obtain a linear increase of data throughput with the number of antennas.

Microcell: A microcell offers a larger deployment footprint than a picocell, such as a residential neighborhood, an office complex, or an entire airport. fjallfoss.fcc.gov/edocs_public/attachmatch/FCC-10-177A1.doc.

NeXt Generation (XG): DARPA's Next Generation (XG) Program goals are to develop both the enabling technologies and system concepts to dynamically redistribute allocated spectrum along with novel waveforms in order to provide dramatic improvements in assured military communications in support of a full range of worldwide deployments. www.darpa.mil/ (search on XG).

Orthogonal frequency division multiple access (OFDMA): like CDMA (above), a system allowing multiple users to share a frequency by encoding signals in distinct ways. Part 15: Alternative name for unlicensed spectrum access, because it conforms to rules set out in 47 C.F.R. 15.

Personal Communications Services (PCS): The Broadband Personal Communications Service (PCS) is in the 1850–1990 MHz spectrum range. The most common use of Broadband PCS spectrum is mobile voice and data services, including cell phone, text messaging, and Internet. www.fcc.gov/encyclopedia/broadband-personal-communications-service-pcs.

Picocell: A picocell offers a wider range of connectivity than a femtocell, but still has a limited range of connectivity and is often employed to provide coverage over an area such as a single floor of a building, a train station platform, or an airport terminal. fjallfoss.fcc.gov/edocs_public/attachmatch/ FCC-10-177A1.doc.

Re-use: A frequency is said to be re-used when it is assigned to different radio stations or systems in different geographical areas; small-cell architectures enable greater re-use than macrocell architectures.

Rural access: Rural and small-town Americans require access to 21st century communications tools and technologies to stay plugged in and competitive in the global economy. Bringing the benefits of mobile broadband to rural America is one the FCC's top priorities. The FCC pursues policies to ensure that consumers in rural areas have access to basic telecommunication services and to encourage the deployment of advanced telecommunication services to rural communities. www.fcc.gov/topic/rural-access.

Shared spectrum access: Includes all situations in which two or more users or wireless applications are authorized to utilize the same range of frequencies on a non-exclusive basis in a defined sharing arrangement, along with any other possibility for multiple users, to access the radio spectrum without exclusive rights. ec.europa.eu/information_society/policy/ecomm/radio_spectrum/_document_storage/studies/shared_use_2012/scf_study_shared_spectrum_access_20120210.pdf.

Small cell: Small cells are low-power wireless access points that operate in licensed spectrum, are operator-managed and feature edge-based intelligence. They provide improved cellular coverage, capacity and applications for homes and enterprises as well as metropolitan and rural public spaces. They include technologies variously described as femtocells, picocells, microcells and metrocells. www.smallcellforum.org/Files/File/SCF-Small_Cells_White_Paper.pdf.

Spatial diversity: Spatial diversity employs multiple antennas, usually with the same characteristics, that are physically separated from one another. Depending upon the expected incidence of the incoming signal, sometimes a space on the order of a wavelength is sufficient. Other times much larger distances are needed. Cellularization or sectorization, for example, is a spatial diversity scheme that can have antennas or base stations miles apart. This is especially beneficial for the mobile communication industry since it allows multiple users to share a limited communication spectrum and avoid co-channel interference. en.wikipedia.org/wiki/Spatial_diversity.

Spectrum: Shorthand for electromagnetic spectrum, the range of frequencies of electromagnetic radiation from zero to infinity. Note: The electromagnetic spectrum was, by custom and practice, formerly divided into 26 alphabetically designated bands. This usage still prevails to some degree. However, the ITU formally recognizes 12 bands, from 30 Hz to 3000 GHz. New bands, from 3 THz to 3000 THz, are under active consideration for recognition. www.atis.org/glossary/definition.aspx?id=7590.

Time division multiple access (TDMA): method for allowing multiple users to transmit and receive at the same frequency by allotting them distinct time slots.

Unlicensed: Unlicensed spectrum refers to radio frequency bands in which technical rules are specified for both the hardware and deployment of radio systems that are open for shared use by an unlimited number of compliant users. The term "unlicensed spectrum" is interpreted to include frequency bands in which the FCC allows sharing with licensed services as well as proposals for possible future unlicensed frequency allocations. Any person or entity may use unlicensed spectrum for either private or public purposes so long as the user's equipment is certified by the FCC and operated in conformity with Part 15 of the Commission's rules. In contrast with most licensed spectrum use, unlicensed spectrum users enjoy no regulatory protection against interference from other licensed or unlicensed users in the band. Although FCC device certification rules and standardized protocols (such as the Wi-Fi Alliance's 802.11 family of protocols) help to mitigate interference, users must accept any interference caused by all compliant devices in the band. www.ntia.doc.gov/files/ntia/meetings/unlicensedspectrum subcommitteereport_01102011.pdf.

White space: Unused spectrum—called white spaces—represents a valuable opportunity for our changing wireless mobile landscape. Sometimes called "wi-fi on steroids,"this spectrum is ripe for innovation and experimental use, holding rich potential for research and commercial purposes. The FCC is moving forward with plans to unlock this spectrum in order to maximize white spaces' value for consumers and businesses. In line with the Commission's duties regarding all spectrum-related actions, the FCC will protect existing spectrum services from possible interference as white spaces innovation grows. www.fcc.gov/topic/white-space.

Wi-Fi: General name for all technologies enabling short-range wireless communication that conform to the IEEE 802.11 family of technical specifications.

WiMAX (Worldwide Interoperability for Microwave Access): Wireless communication technology providing high data transmission rates (tens of megabits per second) over substantial distances (tens of miles). WiMAX is part of the 4G (4th generation) of technology for cellular communications.

ZigBee: Technology specification for wireless communication focused on robust, low-cost, low-power, low-data rate transmissions.

BIBLIOGRAPHY

47 CFR Parts 2 and 95: Additional Spectrum for the Medical Device *Radiocommunication Service.* Federal Register Volume 77, Number 18. Federal Communications Commission, January 27, 2012. www.gpo.gov/ fdsys/pkg/FR-2012-01-27/html/2012-1540.htm.

Allen, E, and H Garlan. "Evolution of Regulatory Standards of Interference." *Proceedings of the IRE* 50, no. 5 (May 1962): 1306–1311.

Allocations, Panel on Frequency, and National Research Council Spectrum Protection for Scientific Uses, Committee on Radio Frequencies. *Handbook of Frequency Allocations and Spectrum Protection for Scientific Uses.* The National Academies Press, 2007. www.nap.edu/ openbook.php?record_id=11719.

Annual Report and Analysis of Competitive Market Conditions With Respect to Mobile Wireless, Including Commercial Mobile Services. Federal Communications Commission, June 27, 2011. hraunfoss.fcc.gov/ edocs_public/attachmatch/FCC-11-103A1.pdf.

An Assessment of the Near-Term Viability of Accommodating Wireless Broadband Systems in the 1675-1710 MHz, 1755-1780 MHz, 3500-3650 MHz, and 4200-4220 MHz, 4380-4400 MHz Bands. U.S. Department of Commerce, National Telecommunications and Information Administration, October 2010. www.ntia.doc.gov/files/ntia/publications/ fasttrackevaluation_11152010.pdf.

An Assessment of the Viability of Accommodating Wireless Broadband in the 1755-1850 MHz Band. U.S. Department of Commerce, National Telecommunications and Information Administration, March 2012. www.ntia.doc.gov/files/ntia/publications/ntia_1755_1850_ mhz_report_march2012.pdf.

Beckman, Richard, Karthik Channakeshava, Fei Huang, V.S. Anil Kumar, Achla Marathe, Madhav V. Marathe, and Guanhong Pei. "Synthesis and Analysis of Spatio-Temporal Spectrum Demand Patterns: A First Principles Approach." *DySpan* (2010).

Benkler, Yochai. "Open Wireless Vs. Licensed Spectrum: Evidence from Market Adoption." *Harvard Journal of Law and Technology* (In press) (November 7, 2011). cyber.law.harvard.edu/publications/2011/ unlicensed _wireless_v_licensed_spectrum.

Benkler, Yochai. "Overcoming Agoraphobia: Building the Commons of the Digitally Networked Environment." *Harv. JL & Tech.* 11 (1997): 287.

Benkler, Yochai. "Some Economics of Wireless Communications." *Harvard Journal of Law & Technology* 16, no. 1 (Fall 2002): 25–83.

Bernstein Research. "The Long View: Myth and Reality About Small Cells and Wi-Fi in Cellular Networks." 2012.

Better Coordination and Enhanced Accountability Needed to Improve Spectrum Management. Telecommunications. Report to Congressional Requesters. U.S. Government Accountability Office, September 2002. www.gpo.gov/fdsys/pkg/ GAO REPORTS-GAO-02-906/pdf/GAOREPORTS-GAO-02-906.pdf.

Block, Fred , and Matthew Keller. "Where Do Innovations Come From? Transformations in the U.S. National Innovation System, 1970-2006." edited by Information Technology and Innovation Foundation, 2008.

Bowers, Raymond, Alfred M. Lee, and Carey Hershey. *Communications for a Mobile Society: An Assessment of New Technology*. Beverly Hills, CA: Sage Publications, 1978.

Brito, Jerry. "The Spectrum Commons in Theory and Practice." *Stanford Technology Law Review* 1 (February 10, 2007). stlr.stanford.edu/ pdf/brito-commons.pdf.

Broadcast Spectrum Incentive Auctions White Paper. CTIA – The Wireless Association and Consumer Electronics Association, February 15, 2011. www.cesweb.org/shared_ files/edm/Press/Spectrum_Whitepaper_FINAL.pdf.

BTR Staff. "Wilmington, Nc, Gets White Spaces Network." *Broadband Technology Report,* 2012.

Buck, Stuart. "Replacing Spectrum Auctions with a Spectrum Commons." *Stanford Technology Law Review* 2 (2002). stlr.stanford.edu/pdf/buck- replacing-spectrum-auctions.pdf.

Bykowsky, Mark M., Kenneth R. Carter, Mark A. Olson, and William W. Sharkey. Enhancing Spectrum's Value Through Market informed Congestion Etiquettes. Federal Communications Commission, Office of Strategic Planning and Policy Analysis, February 2008. hraunfoss.fcc.gov/edocs_public/attachmatch/DOC-280521A1.pdf.

Bykowsky, Mark, Mark Olson, and William Sharkey. *A Market-Based Approach to Establishing Licensing Rules: Licensed Versus Unlicensed Use of Spectrum,* February 2008. hraunfoss.fcc.gov/edocs_public/ attachmatch/DOC-280523A1.pdf.

Bykowsky, Mark, Mark Olson, and William Sharkey. "Efficiency Gains from Using a Market Approach to Spectrum Management." *Information Economics and Policy* 22, no. 1 (March 6, 2010): 73–90.

Calabrese, Michael. "Use it or Share it: Unlocking the Vast Wasteland of Fallow Spectrum," Working Paper, presented at 39th Telecommunications Policy Research Conference (TPRC), George Mason University, September 25, 2011. www.tprcweb.com/images/ stories/2011%20papers/ Calabrese_TPRC_2011.pdf.

Calabrese, Michael. "Ending Spectrum Scarcity: Building on the TV Bands Database to Access Unused Public Airwaves," New America Foundation, Wireless Future Working Paper #25, June 2009. www.newamerica.net/ files/nafmigration/Calabrese_ Working Paper25_EndSpectrumScarcity.pdf.

Calabrese, Michael, Marty Cooper, Mark Crosby, Gary Epstein, Robert Gurss, Mark McHenry, Darrin Mylet, and Jennifer Warren. *Report From the Spectrum Inventory Working Group*. U.S.

Camp, L. Jean. "Designing for Trust." In *Trust, Reputation, and Security: Theories and Practice,* edited by Rino Falcone. Berlin: Springer Verlag, 2003. www.springer.com/ computer/ai/book/978-3-540-00988-7.

Chapin, John, and William Lehr. "Mobile Broadband Growth, Spectrum Scarcity, and Sustainable Competition." Paper presented at the 39th Research Conference on Communications Information and Internet Policy Alexandria, VA, 2011.

Chapin, John, and William Lehr. "The Path to Market Success for Dynamic Spectrum Access Technologies." *IEEE Communications Magazine* (2007).

Chapin, John, and William Lehr. "Time-Limited Leases for Innovative Radios." *IEEE Communications Magazine* (2007).

Chen, Brian X. "Carriers Warn of Crisis in Mobile Spectrum." *The New York Times,* April 17, 2012, sec. Technology. www.nytimes.com/2012/04/18/ technology/mobile-carriers-warn-of-spectrum-crisisothers-see-hyper bole.html.

Chen, Ruiliang, Jung-Min Park, Y. Thomas Hou, and Jeffrey H. Reed. "Toward Secure Distributed Spectrum Sensing in Cognitive Radio Networks." *IEEE Communications Magazine* (2008).

Circular A-11: Preparation, Submission, and Execution of the Budget. Executive Office of the President, Office of Management and Budget, August 2011. www.whitehouse.gov/ sites/default/files/omb/assets/a11_ current_year/a_11_2011.pdf.

Cisco Visual Networking Index: Global Mobile Data Traffic Forecast Update, 2011–2016. Cisco Systems, Inc., February 14, 2012. www.cisco.com/ en/US/solutions/collateral/ ns341/ns525/ns537/ns705/ns827/white_paper_ c11-520862.pdf.

City of Seattle. "Request for Information Public-Private Partnership for the Purpose of Providing Stateof-the-Art Wireless Mission-Critical Voice and Broadband Data Capabilities for Public Safety and General Government." Seattle, WA, 2011.

Clearwire. "Investor Presentation." 2011. files.shareholder.com/downloads/ CLWR/18785270 52x0x448922/c199dd61-e9b1-4694-ab5e-3108146f84 83/Investor%20Presentation%20March%202011%20Final.pdf.

Coase, R. H. "The Federal Communications Commission." *Journal of Law and Economics* 2 (1959): 1–40. "Comments of Shared Spectrum Company, Dynamic Spectrum Use." 2011.

"Comments of the National Telecommunications and Information Administration." no. WT Docket No. 06-150 (2008).

Commerce Spectrum Management Advisory Committee Administration, August 22, 2008. www.ntia.doc.gov/legacy/advisory/spectrum/ meeting_files/081508_csmac_WG3_Report_Revised_ (clean_final).pdf.

"Commercial Spectrum Enhancement Act." In *108-494, 118 Stat. 3986, Title II* 2004.

Committee on Evolution of Untethered Communications, National Research Council. *The Evolution of Untethered Communications.* The National Academies Press, 1997. www.nap.edu/openbook.php?record_id=5968.

Committee on Networked Systems of Embedded Computers, National Research Council. *Embedded, Everywhere: A Research Agenda for Networked Systems of Embedded Computers.* The National Academies Press, 2001. www.nap.edu/openbook.php? record_id=10193.

Committee on Telecommunications Research and Development, National Research Council. *Renewing U.S. Telecommunications Research.* Edited by Robert W. Lucky and Jon Eisenberg. The National Academies Press, 2006. www.nap.edu/openbook.php? record_id=11711.

Committee on Wireless Technology Prospects and Policy Options, National Research Council. *Summary of a Forum on Spectrum Management Policy Reform.* Washington, D.C.: The National Academies Press, 2004. www.nap.edu/catalog.php?record_id=11007.

Committee on Wireless Technology Prospects and Policy Options, National Research Council. *Wireless Technology Prospects and Policy Options.* Washington, D.C.: The National Academies Press, 2011. www.nap.edu/ catalog.php?record_id=13051.

Comprehensive Review of U.S. Spectrum Management with Broad Stakeholder Involvement Is Needed. Telecommunications. Report to Congressional Requesters. U.S. Government Accountability Office, January 2003. www.gao.gov/new.items/d03277.pdf.

The Connected Life: A USD4.5 Trillion Global Impact in 2020. GSM Association and Machina Research, February 2012. connectedlife.gsma. com/wp-content/uploads/ 2012/02/Global_Impact_2012.pdf.

Connecting America: The National Broadband Plan. Federal Communications Commission, March 16, 2010. download.broadband.gov/plan/national- broadband-plan.pdf.

Cooper, Mark. *Efficiency Gains and Consumer Benefits of Unlicensed Access to the Public Airways: The Dramatic Success of Combining Market Principles and Shared Access,* January 2012. www.markcooperresearch. com/SharedSpectrumAnalysis.pdf.

Cooper, Mark. T*he Consumer Benefits of Expanding Shared Use of the Public Airwaves: Liberating long-Term Spectrum Policy from Short-Term Thinking,* November 18, 2011. www.consumerfed.org/pdfs/Consumer- Benefits-of-Shared-Use-Spectrum.pdf.

Cramton, Peter, Evan Kwerel, Gregory Rosston, and Andrzej Skrzypacz. *Using Spectrum Auctions to Enhance Competition in Wireless Services.* Stanford University, Stanford Institute for Economic Policy Research, 2011. www-siepr.stanford.edu/ repec/sip/10-015.pdf.

CSMAC: Report of the Spectrum Management Improvements Working Group. Report on the Second Question. U.S. Department of Commerce, National Telecommunications and Information Administration, Commerce Spectrum Management Advisory Committee, March 1, 2012. www.ntia.doc.gov/files/ntia/meetings/sm_improvements_report_ second_ question.pdf.

Davies, Lewis, and Paul Winter. *Study of Current and Future Receiver Performance: Final Report to OFCOM.* The Technology Partnership (TTP), January 11, 2010. stakeholders.ofcom.org.uk/binaries/research/ technology-research/receiver.pdf.

Defense Business Board. "Report to the Secretary of Defense: A Review of Spectrum Management ", 2010.

Definitions of Efficiency in Spectrum Use. U.S. Department of Commerce, National Telecommunications and Information Administration, Commerce Spectrum Management Advisory Committee, October 1, 2008. www.ntia.doc.gov/legacy/advisory/spectrum/ meeting_files/Spectral_Efficiency_Final.pdf.

Department of Commerce, National Telecommunications and Information Administration, Commerce Spectrum Management Advisory Committee, May 19, 2010. www.ntia.doc.gov/files/ntia/publications/csmac_siwg_ document_0525clean.pdf.

Department for Culture Media and Sport. "Enabling Uk Growth—Releasing Public Spectrum." London, UK, 2011.

Desirable Properties of a Nationwide Public Safety Communication System: Report and Recommendations of the Visiting Committee on Advanced Technology of the National Institute of Standards and Technology. U.S. Department of Commerce, National Institute of Standards and Technology, January 24, 2012. www.nist.gov/director/vcat/upload/ Desirable_Properties_of_a_National_PSN.pdf.

Dyson, George. *Darwin Among the Machines: The Evolution of Global Intelligence.* Helix Books. Reading, MA: Perseus Books, 1998.

Electronic Communications Committee. "Cognitive Radio Systems and Software Defined Radio." www.cept.org/ecc/topics/cognitive-radio -systems-and-software-defined-radio.

Electronic Communications Committee. "Latest Report from Cg Crs." www.cept.org/ecc/groups/ecc/wg-fm/cg-crs/page/latest-report-from-cg-crs

Erpek, Tugba, Mark A. McHenry, and Andrew Stirling. "DSA Operational Parameters with Wireless Microphones." 1–11. IEEE, 2010. www.sharedspectrum.com/papers/.

Faulhaber, Gerald, and David Farber. *Spectrum Management: Property Rights, Markets, and the Commons,* December 2002. assets.wharton. upenn.edu/~faulhabe/SPECTRUM_ MANAGEMENTv51.pdf.

Faulhaber, GR. "The Question of Spectrum: Technology, Management, and Regime Change." *Journal on Telecommunications and High Technology Law* 4, no. 1 (2005): 123–182.

FCC Has Made Some Progress in the Management of Its Enforcement Program but Faces Limitations, and Additional Actions Are Needed, February 2008. www.gao. gov/new.items/d08125.pdf.

FCC. "Report and Order and Further Notice of Proposed Rulemaking." 2009.

FCC TAC, Sharing Working Group. *Spectrum Efficiency Metrics (Draft White Paper – 25 September 2011),* 2011. transition.fcc.gov/oet/tac/tacdocs/ meeting92711/Spectrum_ Efficiency_Metrics_White_Paper_by_TAC_ Sharing_Working_Group_25Sep2011.doc.

FCC Technical Advisory Council Sharing Work Group. "Case Studies: The Role of Receiver Performance in Promoting Efficient Use of the Spectrum." In *Spectrum Efficiency Metrics White Paper,* 2011.

FCC Technological Advisory. "Spectrum Efficiency Metrics of the Fcc Technological Advisory (Draft): Appendix B: Examples of Spectrum Sharing in the Us." 2011.

Federal Communications Commission. "Equipment Authorization (Ea)." transition.fcc.gov/oet/ea/.

Federal Communications Commission. "Mobile Broadband: The Benefits of Additional Spectrum, Omnibus Broadband Initiative." *Technical Paper No. 6* (2010): 2-5.

Federal Communications Commission. "National Broadband Plan: Connecting America." www.broadband.gov/plan/.

Federal Communications Commission. "Technology Advisory Council." www.fcc.gov/encyclopedia/technological-advisory-council.

Federal Communications Commission. "White Space Database Administrators Guide." www.fcc.gov/encyclopedia/white-space-database-administrators-guide.

Feld, Harold. "The Unfinished Radio Revolution: Eight Perspectives on Wireless" Interference Spectrum "Property Rights" and the Doctrine of Adverse Possession." *Journal on Telecommunications and High Technology Law* 9, no. 2 (2011): 523.

Final Report of the Incentives Subcommittee. U.S. Department of Commerce, National Telecommunications and Information Administration, Commerce Spectrum Management Advisory Committee, January 11, 2011. www.ntia.doc.gov/legacy/advisory/ spectrum/reports/Incentives Sub comm_Report_FINAL_01112011.pdf.

Final Report of the Interference and Dynamic Spectrum Access Subcommittee. U.S. Department of Commerce, National Telecommunications and Information Administration, Commerce Spectrum Management Advisory Committee, November 8, 2010. www.ntia.doc.gov/files/ntia/ publications/csmac_interferencecommitteereport_ 01102011.pdf.

Final Report of the Spectrum Transparency Subcommittee. U.S. Department of Commerce, National Telecommunications and Information Administration, Commerce Spectrum

Management Advisory Committee, May 19, 2010. www.ntia.doc.gov/legacy/advisory/spectrum/reports/ CSMAC_STWG_FINAL5182010.pdf.

Final Report of the Unlicensed Uses Subcommittee. U.S. Department of Commerce, National Telecommunications and Information Administration, Commerce Spectrum Management Advisory Committee, January 11, 2011. www.ntia.doc.gov/legacy/advisory/spectrum/reports/UNLICENSED_ USES_REPORT_01112011.pdf.

Forge, Simon, Robert Horvitz, and Colin Blackman. "Perspectives on the Value of Shared Spectrum Access: Final Report for the European Commission." SCF Associates, 2012. ec.europa.eu/information_society/policy/ecomm/radio_spectrum/_document_storage/studies/shared_use_2012/scf_study_shared_spectrum_access_20120210.pdf.

Genachowski, Julius. "FCC Chairman Julius Genachowski Prepared Remarks to International Ctia Wireless 2012." FCC. New Orleans, LA, 2012.

Goodman, Seymour E., and National Research Council Herbert S. Lin, Editors, Committee on Improving Cybersecurity Research in the United States. *Toward a Safer and More Secure Cyberspace.* The National Academies Press, 2007. www.nap.edu/openbook.php?record_id=11925.

"Growth of Mobile Industry Having Positive Economic Impact." TechJounal (2012).

Gruenwald, Juliana. "McDowell: Auctions Will Likely Yield Less Spectrum Than Estimated." *National Journal*, April 17, 2012, sec. Tech Daily Dose. techdailydose.nationaljournal.com/2012/04/mcdowell-auctions-will-likely.php.

Guerrero, Peter F. "Testimony before the Committee on Commerce, Science, and Transportation: History and Current Issues Related to Radio Spectrum Management." edited by United States General Accounting Office, 2002.

Gulyaev, Alexander. "Wapecs: Flexibility Vs. Harmonisation Towards Increasing the Flexilbity of Spectrum Use in Europe." Paper presented at the CEPT Workshop 2011, Copenhagen, DK, 2011.

Hatfield, D. "FCC Regulation of Land Mobile Radio—a Case History." In *Telecommunications: An Interdisciplinary Text,* edited by Leonard Lewin: Artech House Inc, 1984.

Hatfield, Dale N. "Spectrum Issues for the 1990s: New Challenges for Spectrum Management." www.annenberg.northwestern.edu/pubs/ spectrum/.

Hazlett, Thomas. "Liberalizing US Spectrum Allocation." *Telecommunications Policy* 27, no. 7 (2003): 485–499.

Hazlett, Thomas. "The Wireless Craze, The Unlimited Bandwidth Myth, The Spectrum Auction Faux Pas, and the Punchline to Ronald Coase's 'Big Joke': An Essay on Airwave Allocation Policy." *Harvard Journal of Law and Technology* 14, no. 2 (2001). mason.gmu.edu/~thazlett/pubs/The%20 Wireless%20Craze.pdf.

Hazlett, Thomas. "Assigning Property Rights to Radio Spectrum Users: Why Did FCC License Auctions Take 67 Years?" *Journal of Law and Economics* 41, no. 2 (1998): 529–575.

Hazlett, Thomas. "Spectrum Tragedies." *Yale Journal on Regulation* 22, no. 2 (2005): 242. House of Representatives. *Middle Class Tax Relief and Job Creation Act of 2012.* 2nd, H.R. 3630.

Hoven, N, and A Sahai. "Power Scaling for Cognitive Radio." In *Wireless Networks, Communications and Mobile Computing, 2005 International Conference On*, 1:250–255 vol.1. IEEE, 2005. dx.doi.org/10.1109/ WIRLES.2005.1549418.

Information and Communications for Development 2009: Extending Reach and Increasing Impact. The World Bank, 2009. issuu.com/world.bank. publications/docs/ 9780821376058.

ITU Internet Reports 2005: The Internet of Things. Geneva: International Telecommunication Union, November 2005. www.itu.int/dms_pub/itu-s/opb/pol/S-POL-IR.IT-2005-SUM-PDF-E.pdf.

Jackson, GA. "The Early History of Radio Interference." *Journal of the Institution of Electronic and Radio Engineers* 57, no. 6 (1987): 244.

Joiner, Bernard. *Receiver Spectrum Standards: Phase 1 – Summary of Research into Existing Standards.* Boulder, CO: U.S. Department of Commerce, National Telecommunications and Information Administration, Institute for Telecommunication Sciences, July 2003. www.its.bldrdoc.gov/publications/2435.aspx.

Jull, Kenneth, and Stephen Schmidt. "Preventing Harm in Telecommunications Regulation: a New Matrix of Principles and Rules Within the Ex Ante Versus Ex Post Debate." In *Regulation and the Evolution of the Global Telecommunications Industry,* 2010.

Kang, Cecilia. "New iPad Users Slowed by Expensive 4G Network Rates." *The Washington Post,* March 26, 2012, sec. Business. www.washingtonpost.com/business/economy/ new-ipad-users-slowed-by-expensive-4g-network-rates/2012/03/22/gIQARLXYUS_story.html.

Kim, J., V.S. Anil Kumar, A. Marathe, G. Pei, S. Saha, and B. Sunapanasubbiah. "Impact of Geographic Complementarity in Dynamic Spectrum Access." *Proceedings of IEEE International Dynamic Spectrum Access Networks (DySPAN)* (2011).

Krenik, B., and C. Panasik. "The Potential for Unlicensed Wide Area Networks." Wireless Advanced Architectures Group Texas Instruments, 2004.

Krenik, W., and A. Batra. "Cognitive Radio Techniques for Wide Area Networks." Paper presented at the 42nd Design Automation Conference, Anaheim, CA, 2005.

Lasar, Matthew. "FCC Gives Sprint Nextel a Break in 800 Mhz Spectrum Makeover." (2008).

Lehr, William, and John Chapin. "On the Convergence of Wired and Wireless Access Network Architectures." *Internet Economics and Policy* 22 (2010): 33-41.

Machina Research. "The Connected Life: A Usd 4.5 Trillion Global Impact ", 2012.

Machine-to-Machine Communications: Connecting Billions of Devices. OECD Digital Economy Papers. OECD, January 30, 2012. dx.doi.org/10.1787/5k9gsh2gp043-en.

Manual of Regulations and Procedures for Federal Radio Frequency Management. U.S. Department of Commerce, National Telecommunications and Information Administration, May 2011. www.ntia.doc. gov/page/2011/manual-regulations-and-procedures-federal -radio-frequency-management-redbook.

Marcus, Michael J. "New Approaches to Private Sector Sharing of Federal Government Spectrum." *New America Foundation,* no. 26 (2009).

Marcus, Michael J. "Thoughts on Radar/Communications Spectrum Sharing." *Marcus Spectrum Solutions* (2011).

Marshall, Preston. *A Potential Alliance for World-Wide Dynamic Spectrum Access: DSA as an Enabler of National Dynamic Spectrum Management.* New America Foundation, June 2009. www.newamerica.net/files/ nafmigration/Marshall_IssueBrief25_DSA.pdf.

Marshall, Preston. "Extending the Reach of Cognitive Radio." *Proceedings of the IEEE 97*, no. 4 (April 2009): 612 –625.

Matheson, R., and A. Morris. "The Technical Basis for Spectrum Rights." (2007).

McHenry, Mark, Eugene Livsics, Thao Nguyen, and Nivedita Majumdar. "XG Dynamic Spectrum Access Field Test Results." IEEE Communications Magazine 45, no. 6. Topics in Radio Communications (June 2007): 51–57.

McHenry, Mark. *NSF Spectrum Occupancy Measurements Project Summary.* Shared Spectrum Company, August 15, 2005.

McHenry, Mark, and Max Vilimpoc. *Dupont Circle Spectrum Utilization During Peak Hours.* Issue Brief. The New America Foundation and the Shared Spectrum Company, June 20, 2003.

Microsoft. "Cambridge Consortium Completes Successful Trial of Next-Generation Wireless." www.microsoft.com/en-us/news/press/2012/apr12/ 04-25whitespacepr.aspx.

Middle Class Tax Relief and Job Creation Act of 2012, 2012. Middle Class Tax Relief and Job Creation Act of 2012.

Milgrom, Paul, Jonathan Levin, and Assaf Eilat. "The Case for Unlicensed Spectrum." 2011.

MMX Technology LLC. "Spectrum Wiki—the Radio Spectrum Online." www.spectrumwiki.com/Index.aspx.

Mobile Broadband: The Benefits of Additional Spectrum. FCC Staff Technical Paper. Federal Communications Commission, October 2010. download.broadband.gov/plan/fcc-staff-technical-paper-mobile-broad band-benefits-of-additional-spectrum.pdf.

National Research Council. *Spectrum Management for Science in the 21st Century:* National Academies Press, 2010.

National Science Foundation. "Workshop on Enhancing Access to the Radio Spectrum." Arlington, VA, 2010.

National Telecommunications & Information Administration. "500 Mhz Initiative." www.ntia.doc.gov/category/500-mhz-initiative.

National Telecommunications & Information Administration. "An Assessment of the Viability of Accommodating Wireless Broadband in the 1755-1850 Mhz Band." Washington, DC, 2012.

National Telecommunications & Information Administration. "Csmac." www.ntia.doc.gov/ category/csmac.

National Telecommunications & Information Administration. "Interdepartment Radio Advisory Committee (Irac)." www.ntia.doc.gov/ page/interdepartment-radio-advisory-committee-irac.

National Telecommunications & Information Administration. "Irac." www.ntia.doc.gov/ category/irac.

National Telecommunications & Information Administration. *Manual of Regulations and Procedures for Federal Radio Frequency Management (Redbook).* Washington, D.C.2011.

National Telecommunications & Information Administration. "Office of Spectrum Management (Osm)." www.ntia.doc.gov/office/OSM.

National Telecommunications & Information Administration. "Second Interim Progress Report on the Ten-Year Plan and Timetable." Washington, D.C., 2011.

National Telecommunications & Information Administration. "Spectrum Sharing." www.ntia.doc.gov/category/spectrum-sharing.

National Telecommunications & Information Administration. "Spectrum Sharing Innovation Test-Bed Pilot Program." www.ntia.doc.gov/legacy/frnotices/2006/spectrumshare/comments.htm.

National Telecommunications & Information Administration. "Wrc-15." www.ntia.doc.gov/category/wrc-15.

National Telecommunications and Information Administration. "Spectrum Sharing Innovation Test-Bed." Office of the Federal Register, 2008.

Neel, James. "Analysis and Design of Cognitive Radio Networks and Distributed Radio Resource Management Algorithms." Virginia Tech University, 2006.

Newman, T.R., T.C. Clancy, M. McHenry, and J.H. Reed. "Case Study: Security Analysis of a Dynamic Spectrum Access Radio System." In *Global Telecommunications Conference (GLOBECOM 2010), 2010* IEEE, 1 –6, 2010. ieeexplore.ieee.org/stamp/stamp.jsp?tp=&arnumber=5683726 &isnumber=5683069.

NTIA Planning and Processes Need Strengthening to Promote the Efficient Use of Spectrum by Federal Agencies. Spectrum Management. Report to Congressional Committees. U.S. Government Accountability Office, April 2011. www.gao.gov/assets/320/318264.pdf.

OECD. "Enhancing the Competitiveness of Smes in the Global Economy: Strategies and Policies." In *Workshop 2: Local Partnership, Clusters and SME Globalization,* 2000.

Office of Management and Budget. "Circular No. A–11: Preparation, Submission, and Execution of the Budget." edited by Executive Office of the President. Washington, DC, 2011.

Opportunities for Government Adoption of Commercial Technologies. U.S. Department of Commerce, National Telecommunications and Information Administration, Commerce Spectrum Management Advisory Committee, February 2008. www.ntia.doc.gov/legacy/advisory/spectrum/ Commercial_Technologies.pdf.

Opportunities Relating to the Spectrum Sharing Test Bed. U.S. Department of Commerce, National Telecommunications and Information Administration, Commerce Spectrum Management Advisory Committee, December 6, 2007. www.ntia.doc.gov/legacy/advisory/spectrum/ Test_ Bed.pdf.

Ovide, Shira. "At&T/T-Mobile Deal: Explaining Wireless Spectrum." In WSJ Blogs, 2011.

Peha, Jon. "Sharing Spectrum through Spectrum Policy Reform and Cognitive Radio." *Proceedings of the IEEE* 97, no. 4 (2009).

Plan and Timetable to Make Available 500 Megahertz of Spectrum for Wireless Broadband. U.S. Department of Commerce, National Telecommunications and Information Administration, October 2010. www.ntia.doc.gov/files/ntia/publications/tenyearplan_11152010.pdf.

Potential Spectrum Interference Associated with Military Land Mobile Radios. U.S. Government Accountability Office, December 1, 2005. www.gao.gov/products/GAO-06-172R.

"Presidential Memorandum: Unleashing the Wireless Broadband Revolution." The White House, June 28, 2010. www.whitehouse.gov/the-press-office/ presidential-memorandum-unleashing-wireless-broadband-revolution.

Qiang, Christina Zhen-Wei, Carlo Rossotto, and Kaoru Kimura. "Information and Communications for Development 2009: Extending Reach and Increasing Impact." *World Bank issue* (2009).

Rancy, François. "The Need to Harmonize Spectrum for Mobile." Paper presented at the Central and Eastern European Regional Working Group, 2012.

Rappaport, T.S. *Wireless Communications Principles and Practice.* Upper Saddle River, NJ: Prentice Hall, 2002.

Real Wireless. "Report for Ofcom: 4g Capacity Gains Final Report." West Sussex, UK, 2011.

Reed, H., C. Dietrich, J. Gaeddert, K. Kim, R. Menon, L. Morales, and Y. Zhao. "Development of a Cognitive Engine and Analysis of Wran Cognitive Radio Algorithms." In *MPRG Technical Report:* Virginia Tech, 2005.

Reed, J. H., J. T. Bernhard, and J. Park. "Spectrum Access Technologies: The Past, the Present, and the Future." *Proceedings of the IEEE* 100, no. Special Centennial Issue (2012): 1676-84.

Reed, J.H., and C.W. Bostian. "Understanding the Issues in Software Defined Cognitive Radio." Paper presented at the Tutorial for 2005 1st IEEE International Symposium on New Frontiers in Dynamic Spectrum Access Network, Baltimore, MD, 2005.

Recommendations for Improving the Process for Identifying Spectrum for Future Reallocation or Sharing. U.S. Department of Commerce, National Telecommunications and Information

Report From the Spectrum Inventory Working Group. U.S. Department of Commerce, National Telecommunications and Information Administration, Commerce Spectrum Management Advisory Committee, May 19, 2010. www.ntia.doc.gov/legacy/ advisory/spectrum/reports/ CSMAC_SIWG_Document_0525clean.pdf.

Report of the Spectrum Efficiency Working Group. Federal Communications Commission, Spectrum Policy Task Force, November 15, 2002. transition.fcc.gov/sptf/files/SEWG FinalReport_1.pdf.

Report of the Spectrum Policy Task Force. Federal Communications Commission, November 2002. hraunfoss.fcc.gov/edocs_public/ attachmatch/DOC-228542A1.pdf.

Report on Operational Efficiency. U.S. Department of Commerce, National Telecommunications and Information Administration, Commerce Spectrum Management Advisory Committee, 2008. www.ntia.doc.gov/ legacy/advisory/spectrum/meeting_files/ WG2CSMACrevisedjwbntclnfinal.pdf.

Rhodes, Charles. "Interference to Digital Broadband Communications and Spread Spectrum Communications." In *2012 IEEE International Conference on Consumer Electronics (ICCE)*, 283–284. Las Vegas, NV, USA: IEEE, 2012. dx.doi.org/10.1109/ICCE. 2012.6161869.

Rhodes, CW, and GJ Sgrignoli. "Interference Mitigation for Improved DTV Reception." *IEEE Transactions on Consumer Electronics* 51, no. 2 (May 2005): 463–470.

Richards, Ed. "Speech for Dynamic Spectrum Access Forum, Brussels." Ofcom (2012).

Roberson, Dennis A., Cynthia S. Hood, Joseph L. LoCicero, and John T. MacDonald. "Spectral Occupancy and Interference Studies in Support of Cognitive Radio Technology Deployment." In *Networking Technologies for Software Defined Radio Networks,* 26-35: IEEE, 2006.

Rosston, G. L, and J. S Steinberg. "Using Market-based Spectrum Policy to Promote the Public Interest." Fed. Comm. LJ 50 (1997): 87.

Rosston, G. L. "The Long and Winding Road: The FCC Paves the Path with Good Intentions." *Telecommunications Policy* 27, no. 7 (2003): 501–515.

Sanders, Frank, Robert Sole, Brent Bedford, David Franc, and Timothy Pawlowitz. *Effects of RF Interference on Radar Receivers,* n.d. www.its.bldrdoc.gov/publications/06-444.aspx.

Second Interim Progress Report on the Ten-Year Plan and Timetable U.S. Department of Commerce. U.S. Department of Commerce, National Telecommunications and Information Administration, October 2011. www.ntia.doc.gov/files/ntia/publications/ second_interim_progress_report_on_the_ten_year_ plan_and_timetable.pdf.

Shankar, N. Sai, C. Corderio, and K. Challapali. "Spectrum Agile Radios: Utilization and Sensing Architectures." Paper presented at the 1st IEEE International Symposium on New Frontiers in Dynamic Spectrum Access Network, Baltimore, MD, 2005.

Sims, Martin, Michael Newlands, Jonathan Watson, and Dugie Standeford. "Qualcomm and Nokia Propose Authorised Shared Access to Spectrum." In *Policytracker—the spectrum management newsletter,* 2011.

Spectrum Management for the 21st Century: The President's Spectrum Policy Initiative Federal Strategic Spectrum Plan. U.S. Department of Commerce, National Telecommunications and Information Administration, November 2008. www.ntia.doc. gov/legacy/reports/2008/FederalStrategicSpectrumPlan2008.pdf.

Spectrum Management for the 21st Century: The President's Spectrum Policy Initiative Progress Report for Fiscal Year 2007. U.S. Department of Commerce, National Telecommunications and Information Administration, November 2008. www.ntia.doc.gov/legacy/osmhome/ spectrumreform/ FY2007%20Progress%20 Report_ for_Fiscal_Year_2007_Final_25Nov08_rev_1Dec08.pdf.

Spectrum Policy for the 21st Century. U.S. Department of Commerce, National Telecommunications and Information Administration, May 1, 2004. www.ntia.doc. gov/report/ 2004/spectrum-policy-21st-century.

Spectrum Policy for the 21st Century: Improving International Spectrum Management Policies and Framework. U.S. Department of Commerce, National Telecommunications and Information Administration, March 2008. www.ntia.doc.gov/legacy/ osmhome/ spectrumreform/International_ Spectrum_Policy_Improvements_Report3-13-08_Final. pdf.

Spectrum Policy for the 21st Century: Plan to Identify and Implement Incentives That Promote More Efficient and Effective Use of Spectrum. U.S. Department of Commerce, National Telecommunications and Information Administration, November 2008. www.ntia.doc. gov/legacy/osmhome/reports/Incentives_Plan.pdf.

Stine, John A, and Samuel Schmitz. *Model-Based Spectrum Management—Part 1: Modeling and Computation Manual.* MITRE Technical Paper, May 2011. www.mitre.org/work/tech_papers/2011/11_2071/.

Streamlining Federal/Non-Federal Spectrum Sharing. U.S. Department of Commerce, National Telecommunications and Information Administration, Commerce Spectrum Management Advisory Committee, September 19, 2008. www.ntia.doc.gov/ legacy/ advisory/spectrum/ WG2_091308_with_summary_recs.pdf.

Swift, Earl. *The Big Roads: The Untold Story of the Engineers, Visionaries, and Trailblazers Who Created the American Superhighways.* Houghton Mifflin Harcourt, 2011.

Taher, Tanim M., Roger B. Bacchus, Kenneth J. Zdunek, and Dennis A. Roberson. "Long-term Spectral Occupancy Findings in Chicago." In 2011 *IEEE Symposium on New Frontiers in Dynamic Spectrum Access Networks (DySPAN),* 100 –107, 2011. ieeexplore.ieee.org/stamp/ stamp.jsp?tp=&arnumber=5936195.

Thanki, Richard. *The Economic Value Generated by Current and Future Allocations of Unlicensed Spectrum: Final Report.* Perspective Associates, September 8, 2009. apps.fcc.gov/ecfs/document/ view?id= 7020039036.

Thanki, Richard. *The Economic Significance of Licence-Exempt Spectrum to the Future of the Internet.* June, 2012. download.microsoft.com/ download/A/6/1/A61A8BE8-FD55-480B-A06F-F8AC65479C58/ Economic Impact of License Exempt Spectrum - Richard Thanki.pdf.

Transition of Federal Land Mobile Radio Systems to Increase Spectrum Efficiency. U.S. Department of Commerce, National Telecommunications and Information Administration, Commerce Spectrum Management Advisory Committee, December 10, 2008. www.ntia.doc.gov/ legacy/ advisory/spectrum/meeting_files/CSMAC-WG4-2008_ Final_(Dec_10_ 08).pdf.

Transition Report. U.S. Department of Commerce, National Telecommunications and Information Administration, Commerce Spectrum Management Advisory Committee, December 13, 2008. www.ntia.doc.gov/legacy/advisory/spectrum/meeting_files/ CSMAC_Transition_Report_ (121208b_-_CLEAN).pdf.

"U.S. Code Telegraphs, Telephones, and Radiotelegraphs." In Title 47.

The Utility Spectrum Crisis: A Critical Need to Enable Smart Grids. Utilities Telecom Council, January 2009. www.utc.org/fileshare/files/34/Public_ Policy_Issues/Spectrum_ Issues/ finalspectrumcrisisreport0109.pdf.

De Vany, A. S, R. D Eckert, C. J Meyers, and D. J O'Hara. "Property System for Market Allocation of the Electromagnetic Spectrum: A Legal-Economic-Engineering Study, A." *Stan. L. Rev.* 21 (1968): 1499.

Verizon Wireless. "Happy 1st Anniversary, Verizon Wireless 4g Lte!" news.verizonwireless.com/news/2011/12/pr2011-12-05a.html.

Vries, J. Pierre De, and Kaleb August Sieh. "The Unfinished Radio Revolution: Eight Perspectives on Wireless Interference." *Journal on Telecommunications and High Technology Law* 9 (Spring 2011). jthtl.org/content/articles/V9I2/JTHTLv9i2.pdf.

de Vries, Pierre, and Kaleb Sieh. "Reception-oriented Radio Rights: Increasing the Value of Wireless by Explicitly Defining and Delegating Radio Operating Rights." *Telecommunications Policy* (January 2012). dx.doi.org/10.1016/j.telpol.2011.12.009.

Weiser, Phil, and Dale Hatfield. "Policing the Spectrum Commons." *Fordham Law Review* 74 (2005). ssrn.com/abstract=704741.

Weiss, Martin, and Liu Cui. "Spectrum Trading with Interference Rights." In *7th International Conference on Cognitive Radio Oriented Wireless Networks (CrownCom), June 18–20, 2012, Stockholm, Sweden,* 2012. d-scholarship.pitt.edu/11451/.

Werbach, Kevin. "Castle in the Air: A Domain Name System for Spectrum," *Northwestern Univ. Law Rev.,* 104, (2010): 613. papers.ssrn.com/ sol3/papers.cfm? abstract_ id=1557244.

Werbach, Kevin. "Supercommons: Towards a Unified Theory of Wireless Communication." *Texas Law Review* 82 (2004): 863–972.

Zhao, Y., S. Mao, J. Neel, and J. H. Reed. "The Radio Environment Map." In C*ognitive Radio Technology,* edited by B. Fette: Elsevier, 2009.

Zhao, Y., S. Mao, J. O. Neel, and J. H. Reed. "Performance Evaluation of Cognitive Radios: Metrics, Utility Functions, and Methodology." *Proceedings of the IEEE 97,* no. 4 (2009): 642-59.

Zhao, Y., and J.H. Reed. "Radio Environment Map Design and Exploitation." In *MPRG Technical Report:* Virginia Tech, 2005

End Notes

[1] Cisco (2012) *Cisco Visual Networking Index: Global Mobile Data Traffic Forecast Update,* 2011–2016. www.cisco.com/en/US/solutions/collateral/ns341/ns525/ns537/ns705/ns827/white_paper_c11-520862.html.

[2] OECD (2012). *Machine-to-Machine Communications: Connecting Billions of Devices.* OECD Digital Economy Papers, No. 192. dx.doi.org/10.1787/5k9gsh2gp043-en.

[3] GSMA/Machina Research (2012). *The Connected Life: A USD 4.5 trillion global impact in 2020. connectedlife.gsma.com/the*-connected-life-a-usd4-5-trillion-global-impact-in-2020.

[4] Takai, T. (DOD) Information provided by email, June 6, 2012.

[5] Presidential Memorandum (2010). *Unleashing the Wireless Broadband Revolution. www.whitehouse.gov/the*-press-office/presidential-memorandum-unleashing-wireless-broadband-revolution.

[6] NTIA (2012). *An Assessment of the Viability of Accommodating Wireless Broadband in the 1755-1780 MHz Band.* Washington, DC. www.ntia.doc.gov/report/2012/assessment- viability-accommodating-wireless-broadband-1755-1850-mhz-band.

[7] This auction occurred in 2006 and was for the Advanced Wireless Services (AWS) bands (1710-1755 MHz and 2110-2155 MHz).

[8] This factor is calculated in Section 2.2; 1,000 is a conservative estimate.

[9] Swift, E. *The Big Roads.* (2011). New York, NY: Houghton Mifflin Harcourt.

[10] Detailed recommendations are provided in Table ES-1.

[11] Cisco (2012) *Cisco Visual Networking Index: Global Mobile Data Traffic Forecast Update, 2011–2016. www.cisco.com/en/US/solutions/collateral/ns341/ns525/ns537/ns705/ns827/white_paper_c11*-520862.html.

[12] OECD (2012). *Machine-to-Machine Communications: Connecting Billions of Devices.* OECD Digital Economy Papers, No. 192. dx.doi.org/10.1787/5k9gsh2gp043-en.

[13] GSMA/Machina Research (2012). *The Connected Life: A USD4.5 trillion global impact in 2020. connectedlife.gsma.com/the*-connected-life-a-usd4-5-trillion-global-impact-in-2020.

[14] Zhen-Wei Qiang, C., C. Rossotto, and K. Kimura. (2009). "Economic Impacts of Broadband," in *Information and Communications for Development 2009: Extending Reach and Increasing Impact,* World Bank, July 2009, (Figure 3.2). issuu.com/world.bank. publications/docs/9780821376058.

[15] Forge, S., Horvitz, R., and Blackman, C. (2012). P*erspectives on the value of shared spectrum access: Final Report for the European Commission.* SCF Associates. ec.europa.eu/ information_society/policy/ ecomm/radio_spectrum/_document_storage/studies/shared_use_2012/ scf_study_shared_spectrum_access_20120210.pdf.

[16] Hazlett, T.W., R.E. Muñoz, and D.B. Avanzini. (2012). "What Really Matters in Spectrum Allocation Design." Northwestern J. of Technology and Intellectual Property, 10, 93. scholarlycommons.law.northwestern. edu/njtip/vol10/iss3/2/.

[17] Cooper, M. (2012). *Efficiency Gains and Consumer Benefits of Unlicensed Access to the Public Airways. www.markcooperresearch.com/SharedSpectrumAnalysis.pdf.*

[18] Thanki, R. (2012). *The Economic Significance of Licence-Exempt Spectrum to the Future of the Internet.* p.8 download.microsoft.com/download/A/6/1/A61A8BE8-FD55-480B-A06F-F8AC65479C58/Economic Impact

[19] Global System for Mobile Communications, originally *Groupe Spécial Mobile,* was a harmonization of technical wireless standards among European countries.

[20] Vanu Bose, The AIRWIN Project, Vanu, Inc. Whitepaper 1-12, January 2012. www.vanu.com/documents/ technology/airwin-project.

[21] Bernstein Research. (2012). *The Long View: Myth and Reality About Small Cells and Wi-Fi in Cellular Networks.* p. 26 lists the major wireless network equipment vendors as Ericcson, Nokia, Huawei, Alcatel-Lucent, and ZTE.

[22] Presidential Memorandum (2010). *Unleashing the Wireless Broadband Revolution. www.whitehouse.gov/the-press-office/presidential-memorandum-unleashing-wireless-broadband-revolution.*

[23] TechJournal. (Feb. 27, 2012). *Growth of Mobile Industry Having Positive Economic Impact. www.techjournal.org/2012/02/growth-of-mobile-industry-having-positive-economic-impact/.*

[24] Forge, S. *et al.* (2012). op. cit.

[25] The FCC estimates that "mobile data demand is expected to grow between 25 and 50 times current levels within 5 years" and that "the broadband spectrum deficit is likely to approach 300 MHz by 2014." See FCC Staff Technical Paper. (2010). *Mobile Broadband: The Benefits of Additional Spectrum,* pp. 2, 5. download.broadband.gov/plan/ fcc-staff-technical-paper-mobile-broadband-benefits-of-additional-spectrum.pdf.

[26] Ovide, S. (March 21, 2011) "AT&T/T-Mobile Deal: Explaining Wireless Spectrum." *WSJ Blogs: Deal Journal. blogs.wsj.com/deals/2011/03/21/attt-mobile-deal-explaining-wireless-spectrum/.*

[27] NTIA. (2012). *An Assessment of the Viability of Accommodating Wireless Broadband in the 1755-1780 MHz Band. www.ntia.doc.gov/report/2012/assessment-viability-accommo dating-wireless-broadband-1755-1850-mhz-band.*

[28] NTIA. (2010). *An Assessment of the Near-Term Viability of Accommodating Wireless Broadband Systems in the 1675-1710 MHz, 1755-1780 MHz, 3500-3650 MHz, and 4200-4220 MHz, 4380-4400 MHz Band.* See p. 1-6 and figures D45-D55. www.ntia.doc.gov/files/ ntia/publications/fasttrackevaluation_11152010.pdf.

[29] Ofcom. (2012). Speech for Dynamic Access Forum, Brussels. media.ofcom.org.uk/ 2012/03/07/speech-for-dynamic-spectrum-access-forum-brussels/.

[30] Chen, B.X. (Apr. 17. 2012) "Carriers Warn of Crisis in Mobile Spectrum," *New York Times. www.nytimes.com/2012/04/18/technology/mobile-carriers-warn-of-spectrum-crisis-others-see-hyperbole.html.*

[31] Swift, E. (2011). *The Big Roads.* New York, NY: Houghton Mifflin Harcourt.

[32] U.S. Department of Commerce, NTIA, "Second Interim Progress Report on the Ten-Year Plan and Timetable" (Oct. 2011), see particularly Table 2-1, p. 3 and Table 2-3, p. 8. www.ntia.doc.gov/report/2011/second-interim-progress-report-ten-year-plan-and-timetable.

[33] As mentioned above, spectrum from 3550-3650 MHz band was proposed for auction only with large coastal exclusion zones. These would indeed be necessary for high-power, wide-area uses. However, shared use by the kind of low-power uses we propose for the spectrum superhighway would greatly minimize the need for exclusion zones. For more details, see Section 5.1.

[34] 47 U.S.C. § 151, et seq.

[35] 47 U.S.C. § 902(b) (2).

[36] Section 117 of the NTIA Organization Act of 1992, 47 U.S.C. § 927.

[37] The basic function of the IRAC, which has members representing 19 Federal departments and agencies, is to assist the Assistant Secretary in assigning frequencies to U.S. Government radio stations and in developing and executing policies, programs, procedures, and technical criteria pertaining to the allocation, management, and use of the spectrum. www.ntia.doc.gov/page/interdepartment-radio-advisory-committee-irac.

[38] See generally NTIA (2011). *Manual of Regulations and Procedures for Federal Radio Frequency Management (Redbook).* www.ntia.doc.gov/page/2011/manual-regulations-and-procedures-federal-radio-frequency-managementredbook.

[39] Karl Nebbia, Director, NTIA Office of Spectrum Management, presentation to the Commerce Spectrum Management Advisory Committee (CSMAC), Dec. 9, 2009.

[40] 47 U.S.C. §§ 301 and 303.

[41] See Section 2.2 for details of this calculation.

[42] White space is spectrum that is allocated for one use, but has unassigned spectrum in specific locations that can be used for other purposes without impact on the primary usage.

[43] Broadband Technology Report (Jan. 26, 2012). *Wilmington, NC, Gets White Spaces Network. broadbandgear.net/2012/01/wilmington-nc-gets-white-spaces-network/.*

[44] Presentation by Julius Knapp, FCC, IDGA 2012 Conference

[45] 47 C.F.R. parts 2 and 95, Additional Spectrum for the Medical Device Radiocommunication Service. www.gpo.gov/fdsys/pkg/FR-2012-01-27/html/2012-1540.htm.

[46] FCC. (2002). *Spectrum Policy Task Force Report.* hraunfoss.fcc.gov/edocs_public/ attachmatch/DOC-228542A1.pdf.

[47] White Space is spectrum that is allocated for one use, but has unassigned spectrum in specific locations that can be used for other purposes without impact on the primary usage.

[48] "Unlicensed Operation in the TV Broadcast Bands", Second Memorandum Report and Order, 25 FCC Rcd 18661 (2010).

[49] FCC. National Broadband Plan, Ch. 5. *Spectrum.* www.broadband.gov/plan/5-spectrum/ .

[50] FCC. (2010). *Promoting More Efficient Use of Spectrum Through Dynamic Spectrum Use Technologies. www.fcc.gov/document/promoting*-more-efficient-use-spectrum-through-dynamic-spectrum-use-technologies.

[51] National Research Council of the National Academies. (2011). *Wireless Technology Prospects and Policy Options.* Washington, DC: National Academies Press, 2011: www.nap.edu/ catalog.php?record_id=13051.

[52] See, e.g., Bacchus, R.B., K.J. Zdunek, and D.A. Roberson, (2011). "Long-term Spectrum Occupancy Findings in Chicago," in 2011 IEEE Symposium: New Frontiers in Dynamic Spectrum Access Networks. dx.doi.org/10.1109/DYSPAN.2011.5936195; McHenry, M. (2005). "NSF Spectrum Occupancy Measurements: Project Summary," Shared Spectrum Company. www.sharedspectrum.com/measurements/; Mark McHenry, M. and M. Vilimpoc. (2003). "Dupont Circle Spectrum Utilization During Peak Hours: A Collaborative Effort of The New America Foundation and The Shared Spectrum Company," New America Foundation Issue Brief. www.newamerica.net/files/ archive/Doc_File_183_1.pdf.

[53] Like other natural resources, spectrum can be polluted—by spurious emissions from licensed transmitters, for example, but also by unintentional radiation from fluorescent lights, automotive ignition systems, arcing at power line insulators, and many others sources. Protecting the long-term value of spectrum requires attention to such issues, but these concerns are beyond the scope of this report.

[54] Backhaul is the general term for the connections from the core of a network to the outer elements that deliver service. In a cellular phone network, for example, the link from an individual cell tower to the telephone company's core system is part of the backhaul.

[55] Bernstein Research (2012). *The Long View: Myth and Reality About Small Cells and Wi-Fi in Cellular Networks.*

[56] Verizon Wireless News Center (Dec. 5, 2011). *Happy 1st Anniversary, Verizon Wireless 4G LTE!* news.verizonwireless.com/news/2011/12/pr2011-12-05a.html.

[57] FCC TAC; Sunsetting the PSTN; Critical Legacy Transition Working Group; September 27, 2011. transition.fcc.gov/oet/tac/tacdocs/meeting92711/Sept2011_mtg_full.ppt.

[58] NITRD FY2013 Supplement to the President's Budget (2012). p.57 www.nitrd.gov/pubs/2013 supplement/ FY13NITRDSupplement.pdf.

[59] Forge, S. *et al.* (2012). op. cit. p. 29.

[60] Matheson, R. and A. Morris. (2011). *The Technical Basis for Spectrum Rights: Policies to Enhance Market Efficiency.* Washington, DC: Brookings Institution: www.brookings.edu/research/papers/2011/03/03-spectrum-rights-matheson-morris.

[61] "Geolocation" refers to the fact that the database would include the physical location of registered services, as well as other transmission characteristics.

[62] Time Division Multiple Access (TDMA) allows users to share a frequency by transmitting in distinct time slots. Code Division Multiple Access (CDMA) and Orthogonal Frequency Division Multiple Access (OFDMA) allow users to transmit at the same frequency simultaneously but separate their signals by encoding them in distinct ways.

[63] The technique known as spatial diversity refers to the coordinated deployment of multiple antennas in a given area so as to maximize coverage and capability and minimize interference among users sharing the bandwidth.

[64] The most commonly used one is Transmission Control Protocol (TCP), which is one of the core Internet protocols.

[65] Marshall, P. (2009). *A Potential Alliance for World-wide Dynamic Spectrum Access.* New America Foundation Issue Brief #25. www.newamerica.net/files/nafmigration/ Marshall_IssueBrief25_DSA.pdf.

[66] Lasar, M. (June 19, 2008). "FCC gives Sprint Nextel a break in 800 MHz spectrum makeover." *Ars Technica* arstechnica.com/uncategorized/2008/06/sprint-nextel-asks-fcc-for-break-in-800-mhz-spectrum-makeover/; FCC (June 12, 2009). In the Matter of Improving Public Safety Communications in the 800 MHz Band, etc: Report and Order and Order and Further Notice of Proposed Rulemaking. hraunfoss.fcc.gov/edocs_public/ attachmatch/FCC-09-49A1.pdf.

[67] GAO. (2005). *Potential Spectrum Interference Associated with Military Land Mobile Radios. www.gao.gov/products/GAO-06-172R.*

[68] FCC Technical Advisory Council, Sharing Work Group, "Case Studies: The Role of Receiver Performance In Promoting Efficient Use of the Spectrum," Appendix C in *Spectrum Efficiency Metrics White Paper*, Version 1.0, 10 December 2011.

[69] Our view is consistent with that expressed in FCC Interference Immunity Specifications for Radio Receivers, *Notice of Inquiry,* ET Docket No. 03-65 and MM Docket No. 0-39, 18 FCC Rcd 6039 (Mar. 24, 2003), which states "it is not our intent at this time to implement a new regulatory regime that would generally subject all receivers to mandatory standards." hraunfoss.fcc.gov/edocs_public/attachmatch/FCC-03-54A6.pdf.

[70] One of the rare examples: the Minimum Receiver Performance Criteria safe harbor condition for public safety radios, 800 MHz 5th R&O 4th MO&O docket 02-55 (2004) at paragraph 109 ff.

[71] Definition from 47 CFR 2.1 (c): "Harmful Interference. Interference which endangers the functioning of a radionavigation service or of other safety services or seriously degrades, obstructs, or repeatedly interrupts a service operating in accordance with [the ITU] Radio Regulations."

[72] Middle Class Tax Relief and Job Creation Act of 2012, Pub. Law. 112-96. Sec. 6408.

[73] FCC. Comment Deadlines Established Regarding the GPS-LightSquared Technical Working Group www.fcc.gov/document/comment-deadlines-established-regarding-gps-lightsquared-technical-working-group-report

[74] Important exceptions include certain lightly regulated services that are licensed by rule and always accommodated, such as the Family Radio Service (www.fcc.gov/encyclopedia/ family-radio-service-frs), and secondary markets for commercial spectrum, which makes routine leasing possible. There is also the FCC Office of Engineering and Technology Experimental License program, which offers experimental spectrum licenses to entrepreneurs and researchers (transition.fcc.gov/oet/info/filing/elb/).

[75] Benkler, Y. (2011) "Open Wireless v. Licensed Spectrum: Evidence of Market Adoption." cyber.law.harvard.edu/publications/2011/unlicensed_wireless_v_licensed_spectrum.

[76] OECD (2012), *Machine-to-Machine Communications: Connecting Billions of Devices.* OECD Digital Economy Papers, No. 192. dx.doi.org/10.1787/5k9gsh2gp043-en.

[77] GSMA/Machina Research (2012). *The Connected Life: A USD4.5 trillion global impact in 2020. connectedlife.gsma.com/the-connected-life-a-usd4-5-trillion-global-impact-in-2020/.*

[78] Cooper, M. (2012). *Efficiency Gains and Consumer Benefits of Unlicensed Access to the Public Airways. www.markcooperresearch.com/SharedSpectrumAnalysis.pdf.*

[79] Benkler, Y. (2011) op. cit.

[80] FCC. Secondary Markets Initiative. wireless.fcc.gov/licensing/index.htm?job=secondary_ markets. FCC. Spectrum Leasing. wireless.fc

[81] c.gov/licensing/index.htm?job=spectrum_leasing.

[82] Chapin, J.M. and W. H. Lehr. (2007). "Time Limited Leases in Radio Systems", IEEE Communications Magazine, 45, 76-82. dx.doi.org/10.1109/MCOM.2007.374422.

[83] Utilities Telecom Council (2009). *The Utility Spectrum Crisis: A Critical Need to Enable Smart Grids. www.utc.org/fileshare/files/34/Public_Policy_Issues/Spectrum_Issues/ final spectrumcrisisreport0109.pdf.*

[84] FCC Dedicates Spectrum Enabling Medical Body Area Networks. (2012). www.fcc.gov/document/fcc-dedicates-spectrum-enabling-medical-body-area-networks.

[85] A precise number is hard to determine, since many WISPs and RLECs are small entities that do not belong to WISPA or other organizations.

[86] GSMA/Machina Research (2012). *The Connected Life: A USD4.5 trillion global impact in 2020. connectedlife.gsma.com/the-connected-life-a-usd4-5-trillion-global-impact-in-2020/.*

[87] Clearwire Public Investor Presentation, (Feb. 2011) slide 16. files.shareholder.com/downloads/ CLWR/1878527052x0x448922/c199dd61-e9b1-4694-ab5e-3108146f8483/Investor%20 Presentation%20March%202011%20Final.pdf.

[88] Kang, C. (Mar. 22, 2012). Washington Post, "New iPad users slowed by expensive 4G network rates." www.washingtonpost.com/business/economy/new-ipad-users-slowed-by-expensive-4g-network-rates/2012/03/22/ gIQARLXYUS_story.html.

[89] Forge, S. *et al.* (2012). op. cit.

[90] NTIA (2011). *Second Interim Progress Report on the Ten-Year Plan and Timetable.* Table 2-1, p. 3 and Table 2-3, p. 8. www.ntia.doc.gov/report/2011/second-interim-progress-report-ten-year-plan-and-timetable.

[91] transition.fcc.gov/Daily_Releases/Daily_Business/2012/db0508/DOC-313945A1.pdf.

[92] www.ntia.doc.gov/page/interdepartment-radio-advisory-committee-irac.

[93] NTIA (2004). *The President's Spectrum Policy Initiative: Spectrum Policy for the 21st Century. www.ntia.doc.gov/report/2004/spectrum-policy-21st-century.*

[94] The NSTC is the principal means within the executive branch to coordinate science and technology policy across the Federal enterprise. www.whitehouse.gov/administration/ eop/ostp/nstc.

[95] GAO. (2011). *Spectrum Management: NTIA Planning and Processes Need Strengthening to Promote the Efficient Use of Spectrum by Federal Agencies. www.gao.gov/new.items/ d11352.pdf.*

[96] CSMAC. (2012). *Report of the Spectrum Management Improvements Working Group: Report on the Second Question. www.ntia.doc.gov/files/ntia/meetings/sm_improvements_ report_second_question.pdf.*

[97] Middle Class Tax Relief and Job Creation Act of 2012. Title VI - Public Safety Communications and Electromagnetic Spectrum Auctions. www.gpo.gov/fdsys/pkg/BILLS-112hr3630enr/pdf/BILLS-112hr3630enr.pdf.

98 It should be noted that, in addition to incurring fees, a Federal user might also have to bear other costs, such as the costs of meeting protocol compliance requirements and technical standards.

99 Middle Class Tax Relief and Job Creation Act of 2012. Title VI - Public Safety Communications and Electromagnetic Spectrum Auctions. www.gpo.gov/fdsys/pkg/BILLS-112hr3630enr/pdf/BILLS-112hr3630 enr.pdf.

100 Commercial Spectrum Enhancement Act, Pub. L. No. 108-494, 118 Stat. 3986, Title II (2004) (codified in various sections of Title 47 of the United States Code).

101 Middle Class Tax Relief and Job Creation Act of 2012. Title VI - Public Safety Communications and Electromagnetic Spectrum Auctions. www.gpo.gov/fdsys/pkg/BILLS-112hr3630enr/pdf/BILLS-112hr3630enr.pdf.

102 In its ten-year *Plan and Timetable*, the NTIA similarly proposed in 2010 that legislation should "clarify that sharing arrangements are eligible for reimbursements and otherwise liberalize the definition of reimbursable expenses under the CSEA to promote more effective relocation, sharing, and innovative uses of the spectrum..." NTIA (2010). *Plan and Timetable to Make Available 500 Megahertz of Spectrum for Wireless Broadband.* pp. 16-17 www.ntia.doc.gov/files/ntia/publications/tenyearplan_11152010.pdf.

103 Preparation, Submission, and Execution of the Budget. OMB, August 2011. www.whitehouse.gov/sites/ default/files/omb/assets/a11_current_year/a_11_2011.pdf.

104 Block, F. and M. Keller (2008). *Where do Innovations Come From? Transformations in the U.S. National Innovation System, 1970-2006.* Information Technology and Innovation Foundation Report. www.itif.org/files/Where_do_innovations_come_from.pdf.

105 Ofcom. (2012). Speech for Dynamic Access Forum, Brussels. media.ofcom.org.uk/ 2012/03/07/speech-for-dynamic-spectrum-access-forum-brussels/.

106 Forge, S. *et al.*. (2012). op. cit.

107 Forge, S. *et al.* (2012). op. cit. p. 59. According to this report, only the new 450-470 MHz and 2300-2400 MHz bands (120 MHz total) for terrestrial international mobile telecommunications (IMT) are globally harmonized. The challenges in reaching a harmonization agreement on this small portion of spectrum led the authors to recommend a focus on regional harmonization instead.

108 Rancy, F., Director, Radiocommunications Bureau, ITU. (Mar. 29, 2012). *The need to harmonize spectrum for mobile.* (Powerpoint presentation available at www.ceeregionalworkinggroup.net/page.php?99). Rancy notes that the only globally harmonized bands available for 4G to resolve data traffic rapidly exceeding network capacity are the 700/800 MHz and 2.6 GHz bands.

109 www.ntia.doc.gov/files/ntia/publications/international_spectrum_policy_improvements_ report3-13-08_ final. pdf.

110 transition.fcc.gov/ib/sand/agree/.

111 www.fcc.gov/document/major-spectrum-sharing-agreements-canada-and-mexico.

112 NTIA (2011). *Second Interim Progress Report on the Ten-Year Plan and Timetable. www.ntia.doc.gov/files/ntia/* publications/second_interim_progress_report on the ten year plan_and_timetable.pdf.

113 Many of these agreements are listed in the NTIA Manual of Regulations and Procedures for Federal Radio Frequency Management, Chapter 3. www.ntia.doc.gov/page/2011/manual-regulations-and-procedures-federal-radio-frequency-management-redbook.

114 NTIA (undated). *WRC-15. www.ntia.doc.gov/category/wrc-15.*

115 CEPT Electronic Communication Committee (May 18, 2012). *Latest Report from CG CRS. www.cept.org/ecc/groups/ecc/wg-fm/cg-crs/page/latest-report-from-cg-crs.*

116 Policytracker (Mar. 30, 2011). *Qualcomm and Nokia propose authorised shares access to spectrum. policytracker.blogspot.com/2011/03/qualcomm-and-nokia-propose-authorised. html.*

117 CEPT Electronic Communication Committee (undated). Cognitive Radio Systems and Software Defined Radio. www.cept.org/ecc/topics/cognitive-radio-systems-and-software-defined-radio.

118 European Communications Office (Feb 28, 2011) *WAPECS: Flexibility vs. Harmonisation. CEPT Workshop 2011. www.ero.dk/76C3F1A5-55AE-4780-982B-C0690BED3350?frames=no&.*

119 Microsoft News Center (Apr. 25, 2012). *Cambridge Consortium Completes Successful Trail of Next-Generation Wireless. www.microsoft.com/en*-us/news/press/2012/apr12/ 04-25whitespacepr.aspx.

120 Vanu Bose interview April 13, 2012 with Rashmi Dosi, John Leibovitz, and Julius Knapp, all at FCC.

121 NTIA Institute for Telecommunication Sciences (2006). *Effects of RF Interference on Radar Receivers.* NTIA Technical Report TR-06-444. www.its.bldrdoc.gov/publications/ 2481.aspx.

122 Vanu Bose interview April 11, 2012 with Frank Sanders at Boulder ITS.

[123] Trust might be defined as the "intersection of privacy, security and reliability," but there are also the human subtleties of trust perception. (See Camp, L.J. (2003). *Designing for Trust. www.loa.istc.cnr.it/mostro/files/Camp-whatIsTrust.pdf.*

[124] Dyson, G.B. (1998) *Darwin Among the Machines: The Evolution of Global Intelligence.* New York, NY: Perseus Books.

[125] NIST (2012). *Desirable Properties of a Nationwide Public Safety Communication System. www.nist.gov/director/vcat/upload/Desirable_Properties_of_a_National_PSN.pdf.*

[126] OECD (2000). *Enhancing the Competitiveness of SMEs in the Global Economy: Strategies and Policies. Workshop 2: Local Partnership, Clusters and SME Globalization. www.oecd.org/dataoecd/20/5/2010888.pdf.*

[127] Middle Class Tax Relief and Job Creation Act of 2012. Title VI - Public Safety Communications and Electromagnetic Spectrum Auctions. www.gpo.gov/fdsys/pkg/ BILLS-112hr3630enr/pdf/BILLS-112hr 3630enr.pdf.

[128] FCC (May 8, 2012). Prepared Remarks to International CTIA Wireless 2012. transition.fcc.gov/Daily_ Releases/Daily_Business/2012/db0508/DOC-313945A1.pdf.

[129] FCC (2011). *Spectrum Efficiency Metrics* (Technological Advisory Council Sharing Working Group White Paper), Appendix B: Examples of Spectrum Sharing in the US. transition.fcc.gov/oet/tac/ tacdocs/meeting92711/ Spectrum_Efficiency_Metrics_White_Paper_by_TAC_Sharing_Working_Group_25Sep2011.doc.

[130] NTIA (2012). *An Assessment of the Viability of Accommodating Wireless Broadband in the 1755-1850 MHz Band.* Table 2-1, p.6 www.ntia.doc.gov/report/2012/assessment-viability-accommodating-wireless-broadband-1755-1850-mhz-band.

[131] Stine, J. A., and S. Schmitz. (2011). *Model-based Spectrum Management—Part 1: Modeling and Computation Manual.* MITRE Corporation Technical Paper. www.mitre.org/ work/tech_papers/2011/11_2071/.

[132] NTIA (2010). *An Assessment of the Near-Term Viability of Accommodating Wireless Broadband Systems in the 1675-1710 MHz, 1755-1780 MHz, 3500-3650 MHz, and 4200-4220 MHz, 4380-4400 MHz Band. www.ntia.doc.gov/files/ntia/publications/ fasttrackevaluation_11152010.pdf.*

[133] FCC Encyclopedia: White Space Database Administrators Guide. www.fcc.gov/encyclopedia/white-space-database-administrators-guide.

[134] NTIA (2011). Second Interim Progress Report on the Ten-Year Plan and Timetable. ntia.doc.gov/report/2011/second-interim-progress-report-ten-year-plan-and-timetable.

[135] NIST Factsheet: Wireless Innovation Fund. www.nist.gov/public_affairs/factsheet/ wireless_innov2013.cfm.

[136] Middle Class Tax Relief and Job Creation Act of 2012. Title VI - Public Safety Communications and Electromagnetic Spectrum Auctions. www.gpo.gov/fdsys/pkg/BILLS-112hr3630enr/pdf/BILLS-112hr3630enr.pdf.

[137] The metric and discussion presented in this Appendix are adapted from P. F. Marshall, "Scaling, Density, and Decision-Making in Cognitive Wireless Networks." Cambridge University Press (August 2012).

[138] See, e.g., Bacchus, R.B., K.J. Zdunek, and D.A. Roberson, (2011). "Long-term Spectrum Occupancy Findings in Chicago," in 2011 IEEE Symposium: New Frontiers in Dynamic Spectrum Access Networks. dx.doi.org/10.1109/DYSPAN.2011.5936195; McHenry, M. (2205). "NSF Spectrum Occupancy Measurements: Project Summary," Shared Spectrum Company. www.sharedspectrum.com/measurements/; Mark McHenry, M. and M. Vilimpoc. (2003). "Dupont Circle Spectrum Utilization During Peak Hours: A Collaborative Effort of The New America Foundation and The Shared Spectrum Company," New America Foundation Issue Brief. www.newamerica.net/files/ archive/Doc_File_183_1.pdf.

[139] FCC. (2002). *Spectrum Policy Task Force Report.* pp. 3, 4, 14. hraunfoss.fcc.gov/edocs_ public/ attachmatch/DOC-228542A1.pdf.

[140] FCC. (2010). *National Broadband Plan,* p. 95 www.broadband.gov/plan/.

[141] Ibid.

[142] Comments of Shared Spectrum Company, Dynamic Spectrum Use *Notice of Inquiry,* ET Docket No. 10-237, 25 FCC Rcd 13711 (Feb. 28, 2011), at 3. ecfsdocs.fcc.gov/filings/ 2011/02/28/6016171185.html. See also FCC, Dynamic Spectrum Use *Notice of Inquiry,* ET Docket 10-237, 25 FCC Rcd 13711 (Nov. 30, 2010), at 6 & 29. Shared Spectrum Co. is a Virginia-based technology firm that has developed DSA for the military over the past decade, including successful tests in 2006 at Fort A.P. Hill, Virginia, demonstrating DSA policy radios built for DARPA's NeXt Generation ("XG") Communications program. See also McHenry, M. et al. (007). "XG Dynamic Spectrum Access Field Test Results," *IEEE Communications,* 45, pp. 51-57. dx.doi.org/10.1109/MCOM.2007.374432.

[143] FCC. (2002). *Spectrum Policy Task Force Report,* p. 14. hraunfoss.fcc.gov/edocs_public/ attachmatch/DOC-228542A1.pdf.

[144] Comments of Shared Spectrum Company, Dynamic Spectrum Use *Notice of Inquiry,* ET Docket No. 10-237, 25 FCC Rcd 13711 (Feb. 28, 2011), at pp. 4-5. ecfsdocs.fcc.gov/filings/2011/02/28/6016171185.html.

[145] Joiner, B. (2003). *Receiver spectrum standards: Phase 1 – summary of research into existing standards.* NTIA Institute for Telecommunication Sciences Technical Report TR-03-404. www.its.bldrdoc.gov/ publications/ 2435.aspx.

[146] Davies, L., and P. Winter, P. (2010). *OFCOM: Study of Current and Future Receiver Performance.* TTP report. stakeholders.ofcom.org.uk/market-data-research/other/ technology-research/research/spectrum-liberalisation/receiver/.

[147] See FCC. (2002). *Spectrum Policy Task Force Report,* p. 27. hraunfoss.fcc.gov/ edocs_public/attachmatch/DOC-228542A1.pdf.

[148] Unlicensed devices would remain subject to the requirement in 47 CFR 15.5(b) "that interference must be accepted that may be caused by the of an authorized radio station, by another intentional or unintentional radiator, by industrial, scientific and medical (ISM) equipment, or by an incidental radiator."

[149] It is likely that industry standard-setting organizations will develop device performance standards in many cases, as they currently do for cellular systems. See 3GPP: The Mobile Broadband Standard. www.3gpp.org/.

[150] NTIA. Equipment Authorization. transition.fcc.gov/oet/ea/.

[151] NTIA. (2011). Manual of Regulations and Procedures for Federal Radio Frequency Management (Redbook), chapter 10. www.ntia.doc.gov/page/2011/manual-regulations-and-procedures-federal-radio-frequency-management-redbook.

[152] NTIA. Equipment Authorization. transition.fcc.gov/oet/ea/.

[153] RTCA, Inc. (formerly the Radio Technical Commission for Aeronautics) is a private, not-for-profit corporation that develops consensus-based recommendations regarding communications, navigation, surveillance, and air traffic management system issues. RTCA functions as a Federal Advisory Committee. Its recommendations are used by the Federal Aviation Administration (FAA) as the basis for policy, program, and regulatory decisions and by the private sector as the basis for development, investment and other business decisions. www.rtca.org/default.asp.

[154] The European Telecommunications Standards Institute (ETSI) is recognized as an official European Standards Organization by the EU, and generates standards for Information & Communications Technologies including fixed, mobile, radio, broadcast, internet, aeronautical and other areas. www.etsi.org/website/homepage.aspx.

[155] FCC. (2002). *Spectrum Policy Task Force Report.* Section IX. A, recommendation 8. hraunfoss.fcc.gov/edocs_public/attachmatch/DOC-228542A1.pdf.

[156] Feld, H. (2011). "Spectrum "property rights" and the doctrine of adverse possession" in de Vries, J.P. and K. A. Sieh, "The unfinished radio revolution: Eight perspectives on wireless Interference," *Journal on Telecommunications and High Technology Law,* 9 523. jthtl.org/content/articles/V9I2/JTHTLv 9i2_DeVries.PDF.

[157] FCC (2011). *Spectrum Efficiency Metrics* (Technological Advisory Council Sharing Working Group White Paper), Appendix B: Examples of Spectrum Sharing in the US. transition.fcc.gov/oet/tac/tacdocs/meeting 92711/Spectrum_ Efficiency_Metrics_White_Paper_by_TAC_Sharing_Working_Group_25Sep2011.doc.

[158] NTIA (2012). *An Assessment of the Viability of Accommodating Wireless Broadband in the 1755-1850 MHz Band.* Table 2-1, p.6. www.ntia.doc.gov/report/2012/assessment-viability-accommodating-wireless-broadband-1755-1850-mhz-band.

[159] Stine, J. A., and S. Schmitz. (2011). *Model-based Spectrum Management—Part 1: Modeling and Computation Manual.* MITRE Corporation Technical Paper. www.mitre.org/work/ tech_papers/2011/11_2071/.

[160] 47 CFR § 2.102 (f): "The stations of a service shall use frequencies so separated from the limits of a band allocated to that service as not to cause harmful interference to allocated services in immediately adjoining frequency bands."

[161] Federal Register (2008) *NTIA Notice: Spectrum Sharing Innovation Test-Bed. www.federalregister.gov/articles/ 2008/02/05/E8-2050/spectrum-sharing-innovation-test-bed.*

[162] A comprehensive list of experimental facilities can be found at www.nitrd.gov/ subcommittee/wsrd/WSRDTest BedInventory2012.pdf.

[163] Beckman, R. et al. (2010). "Synthesis and Analysis of Spatio-Temporal Spectrum Demand Patterns: A First Principles Approach", in 2010 IEEE Symposium on New Frontiers in Dynamic Spectrum. dx.doi.org/10.1109/DYSPAN.2010.5457859; Kim, J. et al. (2010). "Impact of Geographic Complementarity in Dynamic Spectrum Access." loc. cit. dx.doi.org/10.1109/DYSPAN.2011.5936235.

[164] City of Seattle Request for Information: Public-Private Partnership for the Purpose of Providing State-of-theArt Wireless Mission-Critical Voice and Broadband Data Capabilities for Public Safety and General Government. www.seattle.gov/doit/docs/CityofSeattleRFIfor PublicPrivatePartnership12092011.pdf.

In: Maximizing the Wireless Spectrum ...
Editors: Matthew Echols and Janine Quigley
ISBN: 978-1-62257-941-9

Chapter 2

UNLEASHING THE WIRELESS BROADBAND REVOLUTION*

The White House

PRESIDENTIAL MEMORANDUM FOR THE HEADS OF EXECUTIVE DEPARTMENTS AND AGENCIES

America's future competitiveness and global technology leadership depend, in part, upon the availability of additional spectrum. The world is going wireless, and we must not fall behind. The resurgence of American productivity growth that started in the 1990s largely reflects investments by American companies, the public sector, and citizens in the new communications technologies that are what we know today as the Internet. The Internet, as vital infrastructure, has become central to the daily economic life of almost every American by creating unprecedented opportunities for small businesses and individual entrepreneurs. We are now beginning the next transformation in information technology: the wireless broadband revolution.

Few technological developments hold as much potential to enhance America's economic competitiveness, create jobs, and improve the quality of our lives as wireless high-speed access to the Internet. Innovative new mobile technologies hold the promise for a virtuous cycle -- millions of consumers gain faster access to more services at less cost, spurring innovation, and then a new round of consumers benefit from new services. The wireless revolution has already begun with millions of Americans taking advantage of wireless access to the Internet.

Expanded wireless broadband access will trigger the creation of innovative new businesses, provide cost-effective connections in rural areas, increase productivity, improve public safety, and allow for the development of mobile telemedicine, telework, distance learning, and other new applications that will transform Americans' lives.

* This is an edited, reformatted and augmented version of The White House, Office of the Press Secretary Presidential Memorandum, dated June 28, 2010.

Spectrum and the new technologies it enables also are essential to the Federal Government, which relies on spectrum for important activities, such as emergency communications, national security, law enforcement, aviation, maritime, space communications, and numerous other Federal functions. Spectrum is also critical for many state, local, and tribal government functions. As the wireless broadband revolution unfolds, innovation can enable efficient and imaginative uses of spectrum to maintain and enhance the Government's capabilities.

In order to achieve mobile wireless broadband's full potential, we need an environment where innovation thrives, and where new capabilities also are secure, trustworthy, and provide appropriate safeguards for users' privacy. These characteristics will continue to be important to the adoption of mobile wireless broadband.

This new era in global technology leadership will only happen if there is adequate spectrum available to support the forthcoming myriad of wireless devices, networks, and applications that can drive the new economy. To do so, we can use our American ingenuity to wring abundance from scarcity, by finding ways to use spectrum more efficiently. We can also unlock the value of otherwise underutilized spectrum and open new avenues for spectrum users to derive value through the development of advanced, situation-aware spectrum-sharing technologies.

I therefore am hereby directing that executive departments, agencies, and offices, and strongly encourage that independent agencies, take the following steps:

Section 1. The Secretary of Commerce, working through the National Telecommunications and Information Administration (NTIA), shall:

(a) collaborate with the Federal Communications Commission (FCC) to make available a total of 500 MHz of Federal and nonfederal spectrum over the next 10 years, suitable for both mobile and fixed wireless broadband use. The spectrum must be available to be licensed by the FCC for exclusive use or made available for shared access by commercial and Government users in order to enable licensed or unlicensed wireless broadband technologies to be deployed;

(b) collaborate with the FCC to complete by October 1, 2010, a specific Plan and Timetable for identifying and making available 500 MHz of spectrum as described in subsection (a) of this section. For purposes of successfully implementing any repurposing of existing spectrum in accordance with subsection (a) of this section, the Plan and Timetable must take into account the need to ensure no loss of critical existing and planned Federal, State, local, and tribal government capabilities, the international implications, and the need for appropriate enforcement mechanisms and authorities;

(c) convene the Policy and Plans Steering Group (PPSG) to advise NTIA on achieving the objectives in subsections (a) and (b) of this section. The Secretaries of Defense, the Treasury, Transportation, State, the Interior, Agriculture, Energy, and Homeland Security, the Attorney General, the Administrators of the National Aeronautics and Space Administration (NASA) and the Federal Aviation Administration, the Director of National Intelligence, the Commandant of the United States Coast Guard, and the head of any other executive department or agency that is currently authorized to use spectrum shall participate and cooperate fully, or in the case of independent agencies are strongly encouraged to, in the activities of the Department of Commerce in accomplishing subsections (a) and (b) of this section and promptly provide appropriate funding and staff resources for agency support to these efforts and the work of the PPSG; and

(d) submit, not later than 180 days after the Plan and Timetable described in subsection (b) of this section are completed, to the National Economic Council (NEC), the Office of Management and Budget (OMB), and the Office of Science and Technology Policy (OSTP) an interim report to assess progress against the Plan and Timetable developed in accordance with subsection (b) of this section. Additional interim reports shall be submitted 180 days after the submission of the first interim report and then annually thereafter until such time as the Plan and Timetable are completed. In preparing these reports, the Secretary of Commerce shall work cooperatively with the FCC and other relevant departments, agencies, and offices.

Sec. 2. The Director of OMB shall work with the Secretary of Commerce, through NTIA and in consultation with affected departments, agencies, and offices, to incorporate into the Plan and Timetable referred to in section 1(b) of this memorandum adequate funding, incentives, and assistance to enable executive agencies or other affected entities to accomplish the actions specified in section 1(a) of this memorandum.

Sec. 3. The Secretary of Commerce, working through NTIA, in consultation with the National Institute of Standards and Technology, National Science Foundation (NSF), the Department of Defense, the Department of Justice, NASA, and other agencies as appropriate, shall create and implement a plan to facilitate research, development, experimentation, and testing by researchers to explore innovative spectrum-sharing technologies, including those that are secure and resilient.

Sec. 4. The FCC is strongly encouraged to work closely with the Department of Commerce, through NTIA, to carry out this memorandum as it relates to the FCC, including the repurposing of nonfederal Government spectrum as appropriate and identifying the mechanisms necessary to ensure compliance with the FCC's decisions.

Sec. 5. The NEC, the OMB, and the OSTP (in consultation with the Department of Commerce, working through NTIA, FCC, and the National Security Staff) shall assess, based on the interim report developed pursuant to section 1(d) of this memorandum, whether there has been sufficient progress in achieving the objectives of this memorandum or whether some other mechanism, such as an independent review panel, is needed to address those areas where sufficient progress is not occurring. The NEC, the OMB, and the OSTP shall make any necessary recommendations to the President regarding such progress 45 days after receiving the initial interim report required by section 1(d) of this memorandum and, as appropriate, following subsequent reports.

Sec. 6.

(a) To the extent permitted by law and within existing appropriations, the Department of Commerce, through NTIA, shall provide administrative support for the interagency groups created in this memorandum.

(b) Nothing in this memorandum shall be construed to impair or otherwise affect the functions of the Director of OMB relating to budgetary, administrative, or legislative proposals.

(c) Nothing in this memorandum shall be construed to require the disclosure of classified information, law enforcement sensitive information, or other information that must be protected in the interests of national security.

(d) This memorandum shall be implemented consistent with applicable law and subject to the availability of appropriations.

(e) This memorandum is not intended to, and does not, create any right or benefit, substantive or procedural, enforceable at law or in equity by any party against the United

States, its departments, agencies, or entities, its officers, employees, or agents, or any other person.

Sec. 7. The Secretary of Commerce is authorized and directed to publish this memorandum in the Federal Register.

Barack Obama

INDEX

D

E

F

G

H

N

O

P

T

U

V

W

Y